W9-BYK-957

HOW THE BODY KNOWS ITS MIND

HOW THE BODY KNOWS ITS MIND

The Surprising Power of the Physical Environment to Influence How You Think and Feel

Sian Beilock

ATRIA BOOKS

NEW YORK LONDON TORONTO SYDNEY NEW DELHI

ATRIA BOOKS
A Division of Simon & Schuster, Inc.
1230 Avenue of the Americas
New York, NY 10020

Note to reader: Names and identifying details of some of the people portrayed in this book have been changed. In addition, some individuals and case study details are composites and/or are used illustratively.

First Atria Books hardcover edition January 2015

ATRIA BOOKS and colophon are trademarks of Simon & Schuster, Inc.

For information about special discounts for bulk purchases, please contact Simon & Schuster Special Sales at 1-866-506-1949 or business@simonandschuster.com.

The Simon & Schuster Speakers Bureau can bring authors to your live event. For more information or to book an event contact the Simon & Schuster Speakers Bureau at 1-866-248-3049 or visit our website at www.simonspeakers.com.

Interior design by Kyoko Watanabe
Jacket design by Kathleen Lynch/Black Kat Design
Jacket art by Martin Barraud/plainpicture

Manufactured in the United States of America

10 9 8 7 6 5 4 3 2 1

Library of Congress Cataloging-in-Publication Data

Beilock, Sian.
How the body knows its mind : the surprising power of the physical environment to influence how you think and feel / Sian Beilock.
pages cm
1. Mind and body. I. Title.
BF161.B45 2015
153.7—dc23 2014007653

ISBN 978-1-4516-2668-1
ISBN 978-1-4516-2670-4 (ebook)

To my family

Contents

HOW THE BODY KNOWS ITS MIND

INTRODUCTION

What's Outside Our Head Alters What's Inside

I was running through the woods at full speed when my right foot made contact with a large tree root jutting out of the ground in front of me. My running partner was ahead of me and, thankfully, out of sight, so he didn't see me stumble. We were almost done; only a few more turns on the winding dirt path we'd been following for the better part of five miles and we would be at the car.

I struggled to stay upright, but I couldn't. It was impossible to keep my feet under me. My view of the trees turned sideways and I went down. My hands landed first, followed by my right arm, and then the rest of my body with a loud thud. Everything stopped moving for a few seconds before I could manage to make sure that I was still in one piece. With only a small amount of blood trickling down my leg, I jumped to my feet and was off again. My heart beat loudly in my ears, and my labored breathing reverberated in my chest, but I was moving again and could see my running mate in the distance. A few turns later, the dark greens and browns of the woods gave way to the sun reflecting off the cement parking lot where a lone blue BMW sat parked at the far end. Rolf, my running partner, was already toweling himself off and drinking from a water bottle he had stashed in the car. I had made it. As I walked over to him, I straightened up even more in order to shake off and hide the pain I was in, put on my best smile, and tried to look as confident as possible.

It's always a little nerve-wracking going for a run with someone you don't know, and even more so when your running partner is deciding whether he wants to offer you a job. A week earlier I had received an email from Rolf Zwaan, a professor at Florida State University, where I was about to interview for my first big-time assistant professor position. Rolf asked whether I wanted to go for a run with him in Elinor Klapp-Phipps Park, a nearby nature preserve, the night before my interview started. My plane got in early, so it would be something to do to pass the time, he suggested. To be honest, my first thought was "Absolutely not." Did I really want to spend any more time than I had to with someone who was going to be judging my every move for the next two days? But the more I thought about the idea, the more appealing the run became. Interviewing for a job is a rather sedentary experience—sitting in meeting after meeting all day long—so any opportunity to get in a workout seemed like something I shouldn't pass up. Pumping up my body always seemed to do something positive for my mind, and being outdoors made me feel sharper. As the poet and essayist Ralph Waldo Emerson wrote, "The health of the eye seems to demand a horizon. We are never tired, so long as we can see far enough." And finally, I hoped during our run to learn more about the cool research Rolf had been doing.

Rolf was trying to understand how humans think. I had just read a paper of his in which he argued that we don't "think" the way a computer does, by manipulating abstract symbols in our head. Rather, when we, say, read a story, our brain reactivates traces of previous experiences to make sense of words on a page, almost as if we are mentally simulating being in the story ourselves. In support of his ideas, Rolf and his students had conducted a set of ingenious experiments in which they had people read simple sentences, such as "The eagle was in the sky." The sentence was followed by a picture of an eagle either with wings outstretched (as it would be when it flew in the sky) or by its side (as when perched on its nest). People were asked to indicate if the object in the picture had been mentioned in the preceding sentence. Rolf predicted that, if we understand what we read by mentally placing ourselves in the story, calling upon relevant visual, action, and even emotional information from the past, then we should automatically think about the eagle's shape and respond faster to the eagle that matches the shape implied in the sentence: wings outstretched when we read about an eagle flying, wings down when we read about it in the nest. This is exactly what Rolf found.[1]

Rolf's work points to a new way to think about thinking: it demonstrates that our thinking is *embodied* in that it involves re-experiencing similar bodily experiences from the past. This means that our brain might not make a clear distinction between past memories and what we experience in the present. In other words, our neural hardware might not draw a clear division between thought and action, so that we might be able to use our body and our physical environment to be sharper mentally.

The day after our run in the park, I couldn't help but consider all the factors that would contribute to whether I was going to ace my interview. I realized that lots of influences *outside* my own cortex affected the thought processes that went on *inside* it.

This book is about the many external influences that affect the contents of our mind. The ways my brain worked in my interview, for instance, were influenced by my run the day before. Yes, the exercise made my thoughts sharper, but simply being in the woods had changed my thinking too. And holding my body as if I weren't in pain after my fall actually made me feel better. Whatever we—babies, kids, adults, athletes, actors, CEOs, and you—do from the neck down has a striking impact on what goes on from the neck up. From our brain's standpoint, there isn't much of a line between the physical and the mental. This book explains how we can take advantage of that permeable line and improve our mind by using our body.

Today countless books detail how we think and reason, from Dan Ariely's *Predictably Irrational* to Daniel Kahneman's *Thinking Fast and Slow*. However, very few books consider the influence that our body has on our thinking and decision making, or, more important, how we can leverage our body to change our own mind and the minds of those around us. We tend not to give our body much credit for how we think and feel. But, simply put, kids learn better when they can freely use their body as a tool for acquiring information. For instance, practicing printing letters actually helps kids read. And when you relate mathematical concepts in physical terms, like "Add money to your piggy bank" or "Give away half of your cookies to your sister," kids better understand numbers. The tight relationship between body and mind is also why music and mathematical talent often go hand in hand. Our ability to control our finger movements and our ability to juggle numbers in our head share common neural real estate, which scientists have argued is one reason why building finger dexterity through piano playing can help kids count more fluently in math.

As students' test scores are emphasized more and more, administrators are cutting music, recess, and play in order to keep kids confined to their chairs. But this is a terrible policy, since children learn best through action. How we think is intimately tied to our body and our surroundings. Random hands-on activities cannot make up for our educational woes and our slipping global standing in math and reading skills, but realizing that the body shapes the mind gives us power to structure school to help children learn and think at their best. The current schooling regimen actually hinders children's thinking and learning. In fact our current, sedentary office workplace—and our sedentary lifestyle—keeps adults from thinking and performing at their best too.

The ancient Greeks viewed the human body as a temple that houses the mind. They recognized the linked health of mind and body. By extension, it's also important to pay attention to the environment in which you place your body. I'll tell you the mental power that exercise gives you and show you why body-centered meditation can enhance your ability to focus at work. You will also meet a researcher who has discovered that green space in inner-city projects leads to less violence in the home. And you will learn how to use the power of nature to think more clearly and have more self-control.

Your body helps you learn, understand, and make sense of the world. It can influence and even change your mind—whether or not you are aware of its influence. Companies that make health care products, snacks, and beverages, like Johnson & Johnson and Coca-Cola, have figured this out; they use scientific information about the body's influence to convince us to buy their products. Companies like Google, which understand how important our body is to thinking and creativity, make it easy for their employees to get up and move and to get out and exercise. When your body can move outside the box, your thinking tends to follow.

Your face does a lot more than simply express your emotions; it affects how you register those emotions inside your head and remember them. Frowning and smiling can actually *create* different emotions and attitudes; they're not just the physical *result* of a mood. Standing in a "power pose"—a wide, assertive stance—is linked to increased feelings of power and confidence, which might get you a new client or kudos at work.

Taking Tylenol not only helps ease physical pain; it can also ease the psychological pain of loneliness and rejection. And being physically close to someone else makes us feel more psychologically connected—of one mind. In contrast, being far away sends a subtle signal that we have less common mental ground with others; this is important to know when considering our ever-increasing reliance on virtual communication and whether it is bringing us closer or pushing us farther apart. Why do we gesture when we are on the phone and no one can see us? Can physically manipulating Baoding balls—those little Chinese metal balls that some executives have on their desk—lead to more creative ideas? These are just a few of the questions you will find answers to—answers that have to do with how the body interacts with and responds to its surroundings. Our body has a surprising amount of power in shaping our mind. We just have to learn how to use it.

A few weeks after my interview I got a call from Florida State letting me know I was their second choice; they had offered the job to someone else. I was disappointed (to put it mildly). But my experience, from my mind-clearing run to learning about striking new research on how our body influences our thinking, had convinced me that being successful isn't just dependent on what happens in our head. I realized that what goes on outside our body has a strong influence on the contents of our mind. I had four more interviews to go, and I was determined to use my new knowledge to my advantage. Over the next several weeks I flew to Atlanta to interview at Georgia Tech, Pittsburgh to interview at Carnegie Mellon University, Cincinnati to interview at Miami University, and Greensboro to interview at the University of North Carolina. On each interview I made a point of acknowledging the power my body and my surroundings could have on how I thought and performed. Whether it was a short run the night I flew in, a walk in the park the morning before my first meeting of the day, or simply standing strong and tall during my research talk, I did everything I could to work the line between body and mind.

Of course, anyone who has ever been interviewed for a job knows how idiosyncratic hiring decisions can be; many factors that seem irrelevant to a candidate's performance can play into the final decision. But I am convinced that part of what gave me an edge in these interviews was the fact that I used the power of my body and surroundings to my advantage. I was offered all four jobs.

I hope that, as you accompany me through the rich terrain of body-mind research, the stories I tell will help you too see ways to improve your life and work.

CHAPTER 1

The Laughter Club

THE PHYSICAL NATURE OF EMOTION

It is estimated that one in fifteen American adults, about 21 million, is living with major depression.[1] Most of us feel down in the dumps from time to time, but depression is a never-ending feeling of sadness that affects how you think, how you feel, and how you behave. For people living with a major depressive disorder, everything is gray and life seems bleak, not worth living.

Despite recent headway made in understanding the inner workings of the brain, there is still no treatment for depression that works for everyone. Psychotherapy and drugs like Prozac have helped millions of people stave off depression, but these modalities haven't worked for millions more. The sad fact is that some individuals' depression is resistant to treatment.

Yet consider for a moment that almost all available treatments for depression (whether therapy or medication) target what's going on inside the head. What if there were a way to alleviate depression that went beyond the cortex and altered the body? It might seem odd to focus on the body as an antidote for a disorder seemingly rooted in the mind, but striking new scientific evidence suggests that our body has a powerful influence on our psychological state.

Take the case of Laura, an intelligent, driven twenty-two-year-old. Laura had just graduated from a prestigious Ivy League university and taken her first job at a top public relations firm in Manhattan when her fiancé, Brian, was involved in a car accident and died. Laura was devastated.

Brian and Laura had been high school sweethearts. He was her third kiss and her first love. Even though they had gone off to college on different sides of the country, the two had managed to stay connected as a couple. Brian was her family, the "one," but suddenly, in the midst of planning their late summer wedding and only three short weeks after they had moved into their first apartment together, Brian was gone.

In the several months after Brian's unexpected death, Laura tried to put her life back together. She rented a new apartment in order to get a change of scenery and even went out on a few blind dates her concerned friends had set up for her. But her heart just wasn't in it. While her friends were busy planning their lives, Laura spent her days contemplating the bleakness of life. She constantly broke down in tears and often had trouble getting out of bed, especially on weekends and holidays when she wasn't expected to be anywhere in particular. Her physical energy and her ability to concentrate all but disappeared, and she became increasingly isolated from friends and family. She was noticeably different. As Elizabeth Wurtzel wrote about her own depression in *Prozac Nation*, that's how depression hits: gradually, then suddenly. Laura woke up one morning afraid of what might happen that day, scared to live her life. Everything seemed dark, and she could not think of anything that would make her happy. At some point her mother suggested that she see a psychiatrist, who, not surprisingly, diagnosed Laura with a major depressive disorder.

Laura initially began taking Prozac and going to weekly psychotherapy sessions. At first the drug's effects were almost miraculously positive. Laura couldn't believe how much better she felt. She was more energetic and motivated at work, started seeing her friends, and became interested in life again. Over time, however, she had to take higher and higher doses of Prozac to beat her depression, until the drug seemed to stop working completely. Laura's doctor started her on another medication, but again Laura's depression failed to lift. After a few years she gave up on both drugs and therapy. She was stuck. Then she heard that Botox had been found to help ease depression.

Depressed individuals can often be recognized by their facial expression: a frown with a furrowed brow and downturned mouth. Kurt Cavanaugh, a cosmetic surgeon, immediately picks up on this when patients like Laura walk into his office. On a cool fall day, almost two years to the day after her fiancé had been killed, Laura went to Cavanaugh for Botox treatments.

The active ingredient in Botox is a neurotoxin that paralyzes the muscles into which it's injected. When people get Botox for their frown lines, not only do the frown lines disappear, but their ability to produce unhappy or sour expressions goes away too. Physicians believe that preventing the outward expression of negative emotions helps alter the inner experience of negativity. In other words, certain body movements (or lack thereof) help to change the mind's experience of emotions. On several instances Cavanaugh had casually noted that the moods of his Botox patients after treatments seemed less negative than those of his patients who didn't use Botox. Of course, such differences could easily be due to increased feelings of attractiveness after treatment.

In Hollywood the immense pressure to stay youthful drives actors to use Botox repeatedly. But too much Botox can immobilize your face and your internal feelings. This is bad news for an actor, who needs to convey emotion, but maybe not for someone like Laura with major depression. The media has reported that Nicole Kidman, for example, has had a Botox-induced frozen face; this appeared to be in evidence when she accepted an Academy Award for her performance in *The Hours*. She was clearly crying, yet nothing on her face seemed to be moving. Actors' emotional expressions make their performances more believable to their audience and also help them *internally* experience their character's feelings. The eighteenth-century German philosopher Gotthold Lessing wrote, "I believe that when the actor properly imitates all the external signs and indicators and all the bodily alterations which experience taught him are expressions of a particular [inner] state, the resulting sense impressions will automatically induce a state in his soul that properly accords with his own movements, posture, and vocal tone."[2] Botox can be bad for actors' ability to emote convincingly, yet it can help depressed individuals to quell their internal feelings of sadness by blocking its physical expression.

It may seem odd to think that the expressions we produce outwardly can affect our internal state. After all, we tend to assume that it's the mind that controls the body, not the other way around. But there are direct connections running from the body to the mind. For example, when people are asked to hold a golf tee between their eyebrows in such a way that they have to furrow their brow, they report being in a bad mood.[3] People also judge stories, pictures, and cartoons to be less funny when they are asked to hold a pencil between pursed lips so that their face makes a frown. The opposite is also true: when you hold a pencil in your teeth so that you are smiling, you feel happier. And it's not just facial expressions that send feedback to our brain about our feelings and emotions. When you sit in a slumped position (as opposed to straight, with shoulders back), you don't feel as good about your accomplishments, such as how you just performed on a test or in a presentation. Simply assuming a happy or sad bodily posture, a confident or anxious mien, conveys to our brain what emotional state we are in.

Our facial expressions also affect how we react to stress. Smiling while submerging your hand in ice water for several minutes lessens stress and leads to a quicker recovery from the painful incident than if you don't smile.[4] There really is something to the old adage "Grin and bear it." Of course, there is also a catch: this smile technique works best if you don't know you are doing it—if you form an unconscious smile rather than smile intentionally. In the latter case, the brain seems to catch on and doesn't interpret the bodily expression as happiness. But even faking a smile is better than nothing, because our neural circuitry doesn't always make a clear distinction between what is fake and what is real. Even if you "smile while your heart is breaking," as the ballad suggests, at some level your brain can't help but interpret your smiling as a sign that everything is okay.

A relatively new type of yoga known as Laughter Yoga, or Hasyayoga (*hasya* means "laughter" in Sanskrit), combines laughter with rhythmic breathing. Laughter clubs, where people can engage in this playful activity, have formed from India to Chicago. What starts as forced laughter at some point turns spontaneous and contagious. Laughter not only provides physiological benefits (an abdominal muscle workout and increased lung capacity) but psychological benefits too. Laughter lifts our spirits precisely because our body has a direct line to our mind, telling us how to feel.

In the movie *Mary Poppins,* Uncle Albert (played by Ed Wynn) floats up to the ceiling of his study because he is filled with uncontrollable laughter, singing a song entitled "I Love to Laugh." Uncle Albert's levitation obviously involved some movie fakery, but there is something real in the power of laughter to lighten our moods. A laughing body is an inhospitable host to negativity and stress. There is now even a World Laughter Day—the first Sunday in May, in case you are interested in taking part.

What if your body can't take part in these emotional experiences? This actually happens to the unlucky people born with a rare neurological disorder known as Moebius syndrome. Moebius syndrome prevents people from moving their facial muscles; they can't smile, frown, grimace, or even blink their eyes. It's like "living a life of the mind," one patient said. "I . . . think happy or I think sad, not . . . actually feeling happy or feeling sad."[5] Folks with Moebius syndrome, unable to shape their face into a particular expression, have trouble expressing themselves to others and difficulty experiencing emotions themselves.

To treat Laura's depression, Cavanaugh reasoned that using Botox to prevent frowning might serve as an artificial Moebius syndrome and at least impede negative emotions. The Botox injections he settled on would work on her glabellar frown lines, the wrinkles that occur above the nose and between the eyes and express emotions such as sadness, anger, and distress. Before giving her the injections, however, Cavanaugh asked Laura to complete a common psychological test for evaluating depression, the Beck Depression Inventory,[6] which gauges the severity of symptoms of depression, such as hopelessness and irritability. People taking the test are asked to pick the statements that most closely resemble how they have felt during the previous two weeks. There are twenty-one questions; here is a sample:[7]

UNHAPPINESS

0 I do not feel unhappy.

1 I feel unhappy.

2 I am unhappy.

3 I am so unhappy that I can't stand it.

CHANGES IN ACTIVITY LEVEL

0 I have not experienced any change in activity level.

1 I am somewhat less active than usual.

2 I am a lot less active than usual.

3 I am not active most of the day.

A score of 13 or less signifies that the person is experiencing normal ups and downs (picking mostly 0's and 1's). A score of 29 or more indicates a severe depressive state. Laura scored 42.

In the procedure that followed, which took only a few minutes, Cavanaugh injected Botox into several sites between Laura's eyes and on her forehead. All you have to do is scrunch up your forehead to furrow your brow to see which areas he targeted.

Two months after her Botox treatment, Laura's depression had lifted completely. Given that there was no major change in her life, Cavanaugh's best guess was that her improvement in mood was due to the Botox.

Botox works by blocking the movement of acetylcholine, a neurotransmitter, from the nerves to the muscles. Acetylcholine helps carry signals from the brain to the muscles, letting the muscles know when to tense up. When the flow of acetylcholine is blocked, or at least significantly reduced, the muscle is no longer being told to contract, and so it relaxes. That's why the wrinkled areas into which Botox is injected smooth out and soften: they're not getting the message to tighten. After a while the acetylcholine does get back through. (A normal course of Botox typically lasts between four and six months.) The muscles once again begin to contract and the wrinkles reappear. That's the bad news. The good news is that the wrinkles usually become less prominent after Botox because the muscles have been "trained" to be in a more relaxed state. Perhaps this explains why, when Laura returned to Cavanaugh for a second treatment, her frown lines weren't as pronounced (nor her depressive symptoms as extreme) as on her initial visit. Because Botox can permanently retrain the muscles, the need for further treatment is gradually reduced.

Botox is also approved by the Food and Drug Administration to treat chronic migraines; injections every twelve weeks or so into the head or neck help dull future headaches.[8] Even excessive underarm sweating can be fought with Botox injections to the armpits.[9] Both migraines and sweating have physical as well as emotional triggers. Laura's story suggests that Botox can alleviate depression and improve mental health too, though it's important to point out that Laura knew why she was getting Botox and anticipated that it would help her, just as the Prozac helped initially. But the Botox has kept her depression from returning, so it is unlikely that Laura's changes in mood were simply due to her hopes and expectations about the treatment.

Laura's experience wasn't a fluke. A few years ago a group of psychologists in the United Kingdom tracked down people who had recently had cosmetic treatments. They were particularly interested in comparing the moods of people who had had Botox injections for frown lines (at the same facial sites where Laura had received treatment) with the moods of those who had received other treatments, such as Botox for crow's feet around the eyes, chemical peels, or lip fillers like Restylane. The researchers reasoned that, if not being able to frown makes people happier, then folks who got injections for frown lines should have elevated moods compared to those who got other cosmetic treatments. That is exactly what they found. Limiting negative facial expressions seems to affect mood for the better.[10]

Yet another example of the effectiveness of Botox to change the mind comes from the psychologist David Havas, who specializes in the effects of emotions on how we think and feel. Havas and his colleagues Art Glenberg and Richard Davidson offered first-time Botox users receiving treatment for frown lines a $50 credit toward their treatment if they took part in an experiment before and after the procedure. At both points, the volunteer patients simply had to read a series of sentences depicting positive and negative scenarios:

"You spring up the stairs to your lover's apartment." (happy)
"You open your email in-box on your birthday to find no new emails." (sad)
"Reeling from the fight with that stubborn bigot, you slam the car door." (angry)

Unbeknownst to the volunteers, the researchers measured how long it took them to read the different sentences. Generally it takes longer to read about unfamiliar events than familiar ones, and it also takes longer to read things you don't understand. Reading time thus ends up reflecting how well the information resonates with your own experience—how well you are able to, say, empathize with the emotions you are reading about.

The researchers found that it took the patients roughly the same amount of time to read the happy sentences before and after the Botox treatment. However, they were much slower at reading the sad or angry sentences after the treatment as compared to before. Botox didn't alter comprehension across the board, but it increased the time it took to read and comprehend negative information. According to Havas and his colleagues, this is because Botox prevents people from outwardly and inwardly experiencing the negative situations they are reading about. That's why Botox treatments that prevent people from frowning help to alleviate depression: when you can't form negative facial expressions, you don't feel sad or unhappy thoughts the same way you did before.[11]

How exactly does this facial feedback work? One theory is that, when we read or even think about an emotional event, we mentally relive how we have felt in a similar situation in the past. Put another way, when we see, hear, read, or even think about something bad, we "embody" the experience ourselves. These reactions aren't just in the brain; they extend to our facial expressions and posture. The way we hold our body, in turn, sends signals to the brain about how we feel. That's why when we read a sad story or watch a sad movie, we tend to show evidence of our feelings on our face. But when we aren't able to embody the experience—when there is no feedback from our face to change our mind—emotional processing is hindered. A link in the chain needed for making meaning out of emotional information is missing. For depressed folks who tend to spend a good deal of their time frowning, an inability to furrow their brow to the degree they normally would helps put them in a better mood.

A prolonged inability to form negative facial expressions—a frown or a furrowed brow—actually seems to change how the brain registers negative emotions. People who have had Botox to remove frown lines show reduced activity in neural centers involved in emotion processing. Brain areas such as the amygdala, an almond-shaped region deep inside the brain where negative feelings originate, are less active in people who are asked to mimic

angry facial expressions after Botox as compared to befo
being able to make a sad or angry face for a period of
weeks changes how the brain registers negative emotional e
ences, watering them down, making them less severe.

A recent study conducted in Germany and Switzerland further confirms the ability of Botox to alleviate depressive symptoms. Men and women with an ongoing major depressive disorder were recruited from local psychiatric clinics to get a series of injections in their face (between and just above the eyebrows) over a sixteen-week period. Volunteers knew they might receive injections of either Botox or a placebo, but they didn't know which one. The power of the study comes from the fact that it was double-blind, which means that neither the doctors giving the injections nor the patients themselves knew if they were getting the real Botox injections or a saline solution. The syringes that contained the Botox and the placebo were indistinguishable. But the results were striking. Signs of depression, such as sadness, hopelessness, and feelings of guilt, decreased by an average of 47 percent six weeks after the first treatment for those who actually received the Botox, and the positive benefits remained for the length of the trial. Those in the placebo group didn't show the same marked improvement; their depression held fairly steady across the course of the study.[13]

"Refuse to express a passion, and it dies," wrote the father of modern-day psychology, William James, in 1890.[14] A century later scientists have found support for James's statement in Botox, a drug made popular for its ability to smooth wrinkles.

Facial expressions do not merely express our internal states; they actually affect how emotions are registered in the brain. Charles Darwin was among the first to recognize this body-mind connection. He wrote in *The Expression of Emotion in Man and*

Animals, "The free expression by outward signs of an emotion intensifies it. On the other hand, the repression, as far as this is possible, of all outward signs softens our emotions. He who gives way to violent gestures will increase his rage; he who does not control the signs of fear will experience fear in a greater degree."[15]

The New Science of Embodied Cognition

Darwin argued that the connection of mental states to movement is literally what emotion is (as in the word *emotion* itself), but other philosophers, including René Descartes, thought differently. Descartes claimed that there was a great divide between the mind and the body, that the mind was made up of an entirely different substance than the body. This dualist viewpoint—that our body is irrelevant to understanding how we think, learn, know, and feel—is still widely accepted today. Even many recent brain science books completely overlook the formative role that our body plays in shaping our mind.

The influence of our movements on our thinking and reasoning is only beginning to be measured and appreciated. In the past few years the science of *embodied cognition,* which is in line with Darwin's teachings, has demonstrated how the workings of our mind are entangled with our physical sensations. It sheds new light on the powerful influence that our body has on our mind, and the minds of others. It is providing surprising insights into how our movements influence our decisions and choices, from whom to date to what products to buy. The research on embodied cognition is also changing how we think about how to learn and perform best at school and on the job.

Our mind arises from interactions between our brain, body, and experiences, especially emotional experiences. It's not just that we need the body to show emotions—emotion itself can be traced back to the body. That's why holding a pencil between your teeth in a way that forces you to form a smile puts you in a good mood. It's also why Botox that gets rid of frown lines between the eyes lessens depression. The configuration of your facial muscles sends signals to your brain about how you should feel.

Thinking about the striking connection between body and mind has special significance for me. My career as a cognitive scientist has been heavily influenced by the idea that there is a great divide between the mental and the physical, which has dominated psychology and Western thought for a couple of centuries. This separation of mind and body has been likened to the distinction between the software and the hardware of a computer. I no longer accept the idea that we are simply a set of software programs running on our body hardware because, unlike hardware, our body does influence our mind. As a cognitive scientist, I'm using all the tools available to me to find out how our thinking is shaped by the body, to understand the mind in a larger context, and to find the keys to how we can function at our best.

Accepting the body's influence on the mind helps us make sense of some surprising connections between the physical and the psychological. Take pain as an example. Some of the same brain areas that register physical pain, as when you burn your hand on a hot stove or stub your toe, also log the psychological pain of being rejected by a lover. Because the same neural hardware can serve as a gauge of both mental and physical pain, it makes sense that people who are sensitive to one type of pain (rejection, for example) tend to have more physical complaints.

People who experience the psychological pain of depression also tend to experience a higher rate of physical ailments than those who are mentally healthy.[16]

Body pains also affect our interpretation of psychological pain. Fibromyalgia, characterized by chronic pain and body fatigue, has been linked to loneliness.[17] Likewise, people with chronic pain disorders have a greater tendency to have an "insecure" attachment style, characterized by fear of loneliness and rejection.[18] Enhanced sensitivity to physical pain goes hand in hand with enhanced sensitivity to social pain. Our body has a direct line to our brain and exerts a powerful influence on our mental health and well-being.

Striking new research that my colleagues and I have conducted in my Human Performance Laboratory at the University of Chicago has found evidence of the mind's dependence on the body. For instance, we have discovered that feeling anxiety about doing math is grounded in some of the same folds of brain tissue that register physical pain.[19] My colleagues and I have peered inside the brains of people as they wait to take a math test and discovered that, for those who fear the subject, anticipating doing math looks a lot like being pricked with a needle or burning one's hand on a hot stove. Our mental fears have a lot in common with our physical pains.

It's a standard view that we scientists always conduct a little bit of "me-search" in what we do, and I absolutely want to explore the mind-body connections I have experienced firsthand. As an example, a few months ago I went to pick up my two-year-old daughter, Sarah, from preschool. I immediately noticed she wasn't her happy self, and my maternal alarm signal went off when she asked for some medicine. Was she sick? I checked her forehead, but she didn't feel overly warm, so I asked her what was wrong.

Did her tummy hurt, or maybe she had a sore throat? But neither seemed to be the case. After some more questioning, and a short conversation with one of Sarah's teachers, I got to the bottom of things. Apparently one of the boys in her class had taken a toy from her that she really wanted to play with. He was so mean about not sharing that Sarah had started to cry. Sarah remembered that she took Tylenol when she had a fever and didn't feel well and that taking the medicine usually made her feel better. It was a short leap for her to the idea that the Tylenol would make her feel emotionally better too.

I got to wondering whether Sarah's line of thinking might have some merit, especially since my team had recently discovered that, in the brain, being mentally anxious (say, about doing math) looks a lot like experiencing physical pain. Sure enough, I found research by the husband-and-wife team of Naomi Eisenberger and Matthew Lieberman at UCLA that a daily dose of Tylenol diminishes the hurt feelings that often accompany being socially teased, spurned, or rejected—or getting a toy taken away.[20] Tylenol reduces the sensitivity of the neural circuits involved in pain, so it has the power to lessen both social and physical pain. I wondered if this might work for math-anxious folks too, and, in future research, intend to find out.

Our thinking extends beyond the cortex. My new goal, as both a researcher and a layperson, is to find out just how far this new science of embodied cognition can take us in finding the ingredients we need to function at our best.

CHAPTER 2

Act Early, Think Better Later

Seen from the outside, the Breslin family seems to be living the American dream. John Breslin has a successful orthodontics practice in downtown Chicago, and his wife, Amy, is a stay-at-home mom with a master's degree in early childhood education. They have two beautiful children, Logan, age nine, and Olivia, six. But from very early on, Amy and John had a feeling that there was something not quite right with Olivia.

Amy's pregnancies with both her kids hadn't been easy. She had the typical nausea and tiredness, but her exhaustion went beyond what most women describe. She had looked with amazement at other pregnant women who were excited and bubbly and seemed to float effortlessly down the street, unable to understand how they could feel so good when she was totally worn out. While pregnant with Logan, Amy had been worried that her discomfort might be a sign that something was wrong, but Logan was born at forty weeks, like clockwork, a beautiful, healthy, eight-pound baby boy. His Apgar score, the test used to assess the health of newborns, was a 9 out of 10. One of the nurses told Amy jokingly that only pediatricians' kids get a 10 out of 10. Now

Logan was into sports, the outdoors, and of course anything with a computer screen.

Because of her experience carrying Logan, Amy dismissed the difficulties of her second pregnancy, thinking the discomfort was normal for her. However, a bad case of the flu at sixteen weeks concerned her. The virus put Amy out of commission, giving her a fever of 102 on and off for several days in a row. Her doctor reassured her that the fever wouldn't cause any complications for the baby. But how could she be sure?

After the flu, things went as well as Amy had come to expect. And, to the Breslins' relief, Olivia was born right on time. Their relief was short-lived, however, as it soon became clear that their bright-eyed baby girl was a major handful.

To say that Olivia was colicky, crying for no reason, just doesn't capture the first few months of her life. She cried and cried, a sharp and piercing cry that scared her parents. Olivia didn't like to be put down in her bassinet—or anywhere, for that matter. She had trouble falling asleep, and nothing seemed to make her happy. But Olivia's pediatrician wasn't concerned, saying Olivia was just fussy and would grow out of it.

But when Olivia began missing her motor milestones, there was no way to ignore the signs that something was wrong. Around three or four months of age, when babies are put on "tummy time" and start to lift their head, Olivia just lay there, hardly moving. She couldn't keep her head up very long, as if it was too heavy for her body to support. She didn't start to roll over on her own until she was close to ten months old, and she was slow to sit up and slow to walk. It was as if all the other babies around her age had learned to climb to the top of the jungle gym and Olivia hadn't even figured out that there was a playground.

At some point Olivia's pediatrician tested her for seizures,

strokes, cerebral palsy, and many other syndromes
gist Amy and John had been referred to finally mac
of developmental dyspraxia when Olivia was ar
months old. An impairment in the organization
developmental dyspraxia is usually thought to be due to slow or altered brain development, but its exact causes are unknown.

Amy and John were actually quite relieved when they received Olivia's diagnosis, thinking it wasn't as bad as a terminal disease and would just make her a little slow on the playground. But developmental dyspraxia isn't just about having poor athletic abilities; it's a mind-and-body learning disability. Olivia couldn't hold a crayon, and it took her several tries to open a book. All those activities that other kids learned easily—shoe tying, teeth brushing, using a spoon—Olivia did not. She was also slow to speak and had trouble understanding what others said to her. Motor problems have consequences way beyond being able to catch a baseball. They go hand in hand with mental difficulties too.[1]

About the same time that Olivia's difficulties were convincing her parents of just how closely connected the body and mind are, some four thousand miles across the Atlantic Ocean a group of neuroscientists at the University of Parma in Italy studying monkeys' brains made a discovery that brought them to the same conclusion. The discovery had to do with neurons, the individual nerve cells, in an area of the brain known as the premotor cortex, whose sole function, neuroscientists had assumed for years, was to coordinate bodily movements like reaching for your keys or grabbing a mug of coffee. What the Italian neuroscientists found, however, was that these premotor neurons in primates' brains became excited not only when the monkey moved, such as when he reached for an apple, but also when the monkey saw someone else reach for an apple. Just watching someone else perform an action caused the monkey's motor cortex to fire as if he were doing the action himself.

Professor Giacomo Rizzolatti and his graduate students actually came upon this finding by accident. They had been conducting a typical neurophysiology experiment, recording electrical activity from neurons in a rhesus monkey's brain. The scientists had cut a small hole in the monkey's skull so that they could implant tiny electrodes inside. In this particular study, the tips of the electrodes were placed in individual nerve cells in the premotor cortex. Given that the premotor cortex was known for choreographing movement, the scientists weren't surprised to find that when the monkey reached out and grabbed a peanut and put it in his mouth, the neurons they were measuring turned on. Satisfied with what they were observing, the researchers went to lunch, leaving the monkey wired up in his chair.

When one of the graduate students returned from lunch eating a gelato right in front of the monkey, the electrodes emanating from the animal's cortex signaled that his premotor neurons were firing. The monkey's motor neurons were sensitive to the actions he was observing, even though the monkey himself was completely still![2]

The discovery of these aptly named "mirror neurons," which fire both when an action is produced and when someone is seen performing the same action, point to how our close primate relatives might come to understand the behavior of others.[3] By mentally mirroring the actions he is viewing, as if he were performing them himself, the monkey is able to understand others' goals and intentions. It wasn't such a leap to conclude that humans' ability to decipher other people's actions, intentions, and even feelings might work in a similar way. We understand others by replaying their behavior in our own motor system as if we are performing the behavior ourselves. The existence of mirror neurons means that, for our motor system to properly recognize what's going on around us, it helps to be able to enact the behaviors we observe. This is not such good news for a child such as Olivia, however, given that her developmental dyspraxia hinders her ease of acting. If she can't fire up her own motor system to produce fluent bodily movements, she's likely to have trouble understanding the actions and intentions of others.

For years brain scientists have assumed, like Descartes, that mind and body are largely separate entities. But the discovery of mirror neurons in the monkey's premotor cortex paints an entirely different picture of the mind-body connection. Instead of seeing the body as a passive vehicle that the mind puts to use, we now realize that our body and its experience acting in the world influence the contents of our mind in unexpected ways. Being able to perform an activity—whether it's eating, brushing our hair, or throwing a ball—gives us the ability to recognize what others are doing and, more important, why they are doing it.

Of course, other twentieth-century scientists had begun looking at the body's intelligence before the mirror neuron discovery. In the 1960s the Swiss philosopher and psychologist Jean Piaget argued that body movements served as a foundation of knowledge.[4] Piaget believed that babies possess "sensorimotor intelligence," which means that their actions help them form ideas about the world. Infants don't differ from adults simply because they have less knowledge or lower brain-processing power, Piaget pointed out, but because they haven't yet spent as much time interacting with their surroundings. The contents of infants' thoughts are actually different from those of adults. Einstein said that Piaget's idea that children have their own special kind of logic is "so simple that only a genius could have thought of it."[5]

One of Piaget's insights about the strong connection between moving and understanding came when he saw his seven-month-old daughter, Jacqueline, drop a plastic duck she was holding onto the blanket she was playing on. The toy landed in a fold in the blanket so she could no longer see it. Jacqueline had seen the toy drop, and it was still within her tiny arm's reach, but she made no attempt to recover it. Curious, Piaget put the duck in front of her; then, just as she was about to grasp it, he slowly moved it from her view. Though she clearly saw him hide the duck, she made no attempt to retrieve it. Jacqueline was quite captivated by the duck, but the instant the toy disappeared, she acted as if it had never existed. Out of sight really meant out of mind.[6]

From his interactions with Jacqueline and his observations of other children, Piaget concluded that infants don't understand that objects continue to exist when they themselves can't see them. Piaget believed that children learn this notion of "object permanence" only as they gain experience interacting with the world themselves. Though more recent research has poked holes in some of Piaget's claims, [7] his belief in the power of action was spot on. Our actions help cue our minds about how the world works and why people tend to act in particular ways.

A typically developing toddler travels about forty-seven football fields a day and accumulates an average of seventeen falls an hour,[8] giving him a ton of experience navigating the world. It's pretty easy to disregard crawling as insignificant, but it is hugely important to a baby's physical and mental growth. One reason is that crawling is not such an easy thing to do. As Steve Pinker writes in *The Language Instinct*, the motor abilities of children—whether crawling, walking, or even grabbing a pencil—are "some of the hardest engineering problems ever conceived."[9] We can teach a computer to play chess against the greatest minds of our time, but getting one to walk or crawl as effectively as a human child is still a challenge. More important, the movements of a toddler tell us a lot about how physical coordination relates to mental dexterity.

Take an experimental setup known as "the visual cliff." It's the baby version of base jumping, where people jump off a cliff with only a parachute on their back. Of course, the baby doesn't have a parachute and there isn't a real cliff, but the infant doesn't know this. Here's how it goes: A baby is placed on a large, plexiglass-topped table. Half of the table has a checkerboard pattern just underneath the surface, making it appear safe to crawl on. But the other half of the plexiglass is clear, giving the illusion that the tabletop falls away in a "visual cliff." It's perfectly safe for the baby to proceed, but the baby isn't so sure. The problem from the baby's perspective is that there is a cool toy on the other side of the visual cliff that she desperately wants. What's a baby to do?

Some babies avoid the visual cliff, while others crawl heedlessly forward. Who are these little risk-takers, and what makes them so different from their more cautious counterparts? As it happens, the infants who crawl straight over the apparent cliff are the more inexperienced crawlers. Babies who have been crawling a lot longer avoid the cliff; their experienced motor system emits signals warning that the drop-off might not be safe.[10]

Interestingly, even babies who avoid the visual cliff when crawling often wheel right over the edge without a moment's hesitation when they are in a walker that allows them to scoot around with their feet on the floor.[11] They are expert crawlers, but not expert walkers, so their motor system doesn't send out a warning that *walking* over the edge of the cliff won't be safe. This is one reason why baby walkers are so dangerous: they allow mobility beyond babies' bodily capabilities, and, as a result, the infants don't learn to predict the outcome of their actions. Babies in walkers tend to go right over visual cliffs—like the stairs in their house.

In the mid-1990s, when baby walkers were at the height of their popularity in the United States, the Consumer Products Safety Commission reported that walkers were responsible for more injuries (broken bones, chipped teeth, head injuries, and others) than any other baby product in circulation. In 2004 Canada banned baby walkers; possession can lead to fines up to $100,000 or six months in jail.[12] Not only are walkers dangerous, but they actually retard motor development. Babies who spend a lot of time in walkers don't learn to stand by themselves as soon as they normally would and have a harder time walking unaided because they are used to having their weight supported by the device. Indeed, delays caused by the baby walker are striking: every twenty-four hours of total walker use is associated with delays in walking alone of about three days and delays in standing alone of almost four days.[13]

Diapers can also hinder motor development. Walking is difficult for babies, but it's even harder when you have to wear a bulky diaper between your legs. Old-fashioned cloth diapers are especially disastrous for walking because the bulk causes babies to take wider, bowlegged steps, but even modern disposable diapers that are designed to be thin and light can adversely affect gait. When wearing a diaper, babies fall more and look more awkward when walking.[14] When infants are naked, they walk better. Yet we hardly give them any time to run around in their birthday suit. In one study of diaper use, infants just over a year old averaged only forty-one minutes of naked walking per week; a third of the babies never walked naked.

How babies move affects their cognitive functioning too. Nine-month-old infants who can crawl have better memory than their same-age counterparts who don't yet locomote freely on their own.[15] The more infants explore their surroundings, the more practice they get using their memory about one situation to guide their actions in new surroundings. Continuously flexing their mind in this way endows infants with sharper thinking skills. In contrast, baby walkers have been linked to delays in hitting cognitive milestones, such as interacting with a caregiver and understanding the thoughts and intentions of others. These delays in mental development are still present almost a year after walker use has ceased.[16]

Information doesn't travel in only one direction, from thought to action; action also creates thought. Not only do babies come to understand by experience how objects work and where it's safe to walk, but their mental skills, such as understanding others' intentions, thoughts, and feelings, also come from being able to act on their own in the world.

Simply put, babies understand others' intentions much better when they can do what they are seeing others do. Think about reaching out for an object: what exactly we reach for tells others something about our intention. Do we want to pick up a book or a stuffed bear or a ball? If these toys are all in the same toy box, it's basically the same movement to reach each one, but the intention is different. Babies come to realize this only when they can pick up toys themselves. It seems obvious to you and me, but not to a three-month-old. Babies who can't yet reach out and grab objects on their own aren't as adept at noticing when a person switches from grabbing one toy to grabbing another. Babies need the opportunity to pick up the toys themselves. Wearing a pair of "sticky mittens," which have Velcro on the palm, lets them easily pick up toys by swiping or batting at them. All of a sudden they start to notice when someone picks up a new toy.[17] Like a light switch shifting from the "off" to the "on" position, infants' experience of reaching and grasping for toys infuses their tiny premotor cortex with the ability to notice someone else's goals. As with the rhesus monkeys in Rizzolatti's experiments, the human babies are able to mirror the actions of others and understand others' intentions because they have experience interacting with the toys themselves. This is why Olivia Breslin, the six-year-old with developmental dyspraxia, is having a hard time understanding what others are doing: she herself can't do the things she is observing.

Some famous people have been diagnosed with developmental dyspraxia and have recounted the difficulties this motor disorder can cause. Daniel Radcliffe, the British actor who plays Harry Potter in the movie version of the popular book series, apparently suffers from dyspraxia and still has trouble tying his shoes. "I sometimes think, why, oh why, has Velcro not taken off?" Radcliffe jokes. He said of his school days, "I [had] a hard time at school, in terms of being crap at everything, with no discernible talent." Fortunately he's found his niche, but he struggled with the basics: writing and math.[18] Motor difficulties lead to all sorts of mental difficulties, especially in the classroom.

The impact of physical development on intelligence is made clear in a recent study conducted by researchers at the National Institute of Child Health and Human Development in Washington, DC. The team, led by the psychologist Marc Bornstein, followed 374 infants from five months of age through adolescence, periodically assessing their intelligence and achievement. The researchers' findings were striking. The actions kids could perform at five months predicted not only their IQ at four and ten years of age but their academic achievement (reading comprehension and math problem-solving) at age fourteen. These actions included "tummy time," when infants could lift their head and shoulders for several seconds at a time; when they could sit by themselves; and how often they attempted to reach out and grab the objects around them. The researchers were able to show that the link from action to thought was explained not by the parents' intelligence or education level but by the infants' physical capabilities. When kids can sit up by themselves, their hands are free to reach out and grab objects, which allows them to learn things about the world that they wouldn't otherwise. Infants learned that their actions could change their environment, which helped shape their understanding of others' actions and intentions. Even the language adults used around moving infants tended to be more complex, something known to enhance infant cognitive development. In short, action and intelligence are intertwined. The end result, Bornstein says, is that "motor-exploratory competence in infancy is a catalyst for adolescent academic achievement."[19]

The link between acting and thinking can be seen in all sorts of activities. Fast-forward from five-month-olds to preschoolers. Most four- and five-year-olds can sing the alphabet song and print their name, but few can actually read. What does it take to push these kids to accomplish this cognitive milestone? Practice in naming letters and sounds out loud is part of it, but it's not the whole story, or perhaps even the most important one. Practice in *printing* letters is imperative to reading success: when the body figures out how to write letters, the mind follows suit in being able to read them.

Karen James, a neuroscientist at Indiana University, found that preschool children who took part in a month-long reading program where they practiced *printing* words were better at letter recognition than kids who did the same reading program but practiced *naming* the words instead. Letter recognition isn't enhanced as much by reading the letters as it is by printing them.[20]

James thinks that the reason printing practice is so important for letter recognition and, ultimately, for reading success lies in a fold of tissue near the bottom of the brain that is part of the human visual system, called the fusiform gyrus. The fusiform gyrus is where letters are known to be processed in the brains of adults. Brain imaging studies have shown that the left fusiform responds more strongly when English-speaking adults see individual English letters as opposed to Chinese characters. Scientists often assume that this letter specialization stems from our extensive reading experience, but James thinks that writing experience is the reason. After preschoolers took part in the month-long reading program, their left fusiform gyrus really tuned in to the letters. Most important, this letter sensitivity was much more apparent in the children who learned to print the letters than in those who only read them. That is, the brain area involved in recognizing letters didn't fully engage until kids learned to produce the letters themselves.

James's findings may explain why children diagnosed with dyslexia are often also delayed in their motor development. We often think of dyslexia as simply the tendency to confuse or transpose letters, for example, mistaking "b" for "d." But dyslexia is a reading disorder that affects people's ability to recognize letters and to separate the sounds that make up spoken words. If actually printing letters helps the brain recognize them, then the motor difficulties that dyslexic kids experience might play a big role in their letter-learning ability. When people can't act, they have a hard time understanding.

Such body-mind connections once puzzled scientists, but now the link makes sense. Even though reading seems to be an activity entirely confined to the brain, it also involves the body. And since printing practice helps jump-start areas of the brain needed for letter identification, it is not hard to imagine all sorts of other ways in which motor experience can change the brain. In short, we learn by doing.

From Music to Math

The Breslins tried everything they could to help their daughter cope with her developmental dyspraxia. By the time Olivia was a toddler she was going to twice-weekly occupational therapy sessions, where she learned to balance on medicine balls, put her coat on a hook, and throw a ball. She also had speech therapy sessions to help her move her mouth and lips to make specific sounds and speak more clearly. Olivia has definitely shown signs of progress and, at six, is in a kindergarten program. She is still behind in her motor development, but at least she is able to engage enough in class activities to make it through half a day with her typically developing friends.

Olivia also takes piano lessons. It was her mom's idea, after seeing how much Olivia enjoyed banging on the piano they had in their living room. Curiously, about eight months into these weekly lessons, Olivia's performance in school markedly changed, specifically in math. Her ability to count improved dramatically, and her basic grasp of what numbers meant showed marked improvement too. Her parents wondered if there was a link between her piano playing and her math proficiency.

Others have wondered about a connection between music and math or even music and thinking power, often called "the Mozart effect." A research finding in the early 1990s purports that listening to Mozart improves IQ.[21] That discovery has since been used to support the idea that playing a Mozart opera like *The Magic Flute* through headphones placed on a pregnant woman's belly will enhance the chance that the developing fetus will get into Harvard. A quick Google search of "Mozart effect" yields CDs, DVDs, and books detailing how classical music will make your child smarter. Mozart's music has been credited with everything from boosting the milk production of cows to helping to break down waste at sewage plants.[22] The former governor of Georgia Zell Miller even proposed including in the state budget $105,000 a year to provide every child born in Georgia with a tape or CD of classical music.[23] Tennessee followed in Georgia's footsteps. Eventually a small cottage industry of Mozart CDs for toddlers, babies, and developing fetuses sprang up.

Unfortunately it doesn't look like there really is a Mozart effect. When scientists analyzed the results of almost two dozen studies on the Mozart effect, the benefit to IQ performance was too small to be significant. It certainly doesn't hurt your child to listen to classical music, but it doesn't seem to make him any smarter.[24] Indeed the title of a recent paper from a group of psychologists at the University of Vienna pretty much sums it up: "Mozart Effect, Schmozart Effect."[25] Scientists aren't convinced that even the small intelligence benefits of listening to Mozart that are sometimes found are due to the music itself. Mozart's music is quite stimulating to neurons, and such excitement is generally registered in the right hemisphere of the brain, which supports many of the reasoning abilities researchers have tested in their search for a link between music and thinking power. Perhaps the Mozart effect that has been found is really just about being aroused or excited. In support of this excitement idea, it's been shown that just listening to a passage from a scary Stephen King novel also enhances people's performance on common IQ tests—especially if they really get into the story.[26]

Though the claims that listening to Mozart can make you smarter are overstated, there are tons of cases of kids excelling in both music and school. Consider the daughters of Amy Chua, author of the best-selling book *Battle Hymn of the Tiger Mother*, which details her strict upbringing of Sofia and Louisa (Lulu). Chua wouldn't allow her children to attend sleepovers, watch TV, or play video games because she felt their time was better spent concentrating on academics and playing piano and violin. By American standards, Chua's mandates that her girls practice their instruments for several hours after doing extra academic work (especially in math) might seem overly demanding, but her methods produced two daughters who excel in both music and math.

The MATHCOUNTS competition, a national middle school mathematics program that promotes mathematics achievement through exciting and engaging spelling bee–type contests,[27] regularly has winners proficient in both math and music. Students solve problems such as "If Kenton walks for 60 minutes at the rate of 3 mph and then runs for 15 minutes at the rate of 8 mph, how many miles will he travel?" (The answer is five miles.) All the members of the first place–winning team of the 2011 Los Angeles County Chapter MATHCOUNTS Competition, besides being math whizzes, play a musical instrument.[28]

Why might musical training go hand in hand with enhanced mathematics skills? It all comes back to the body. In the past several years, scientists have tuned in to the link between our ability to control our fingers (which is usually highly developed in musicians) and mathematical performance. Fingers and numbers share common neural real estate in the brain; the parietal cortex in particular is involved in both.[29] Recent work shows that practicing with the body in music training helps kids develop their brain for math. The opposite is true too—several cases have occurred over the years of people who suddenly lose their ability to use their fingers properly and also to juggle numbers in their head.[30]

Take Henry Polish, who, at fifty-nine, woke up one day unable to do a simple arithmetic calculation or dial a phone number. Henry worked as an insurance agent at a small firm in Atlanta, Georgia, and was accustomed to performing mathematical calculations on a daily basis. So you can imagine his surprise when he sat down to pay some bills one Saturday morning after breakfast and found that he could no longer add a series of single-digit numbers in his head. He was alert, speaking fine, and had no vision problems. He couldn't figure out what was going on. His wife suggested they go to the local emergency room, but he thought it would be best if he first called a friend of theirs who was a doctor. When Henry couldn't remember the phone number, though, he relented, and his wife took him to the ER.

Doctors did a complete neurological workup on Henry and found a very strange pattern: he could speak and understand, move and follow directions, but he had trouble with activities involving his fingers and numbers. When the neurologist asked Henry to link his two pinky fingers together, for instance, he couldn't do it. He just sat there, dumbfounded at his inability to coordinate simple movements of his two hands. He understood the instructions and knew what his hands were supposed to do, but they just wouldn't cooperate. The doctor then asked Henry to close his eyes and began touching his fingers, one by one, asking him which finger he had just touched. Henry's answers were no more accurate than if he were guessing at random. Henry also found it difficult to recognize simple Arabic numerals (for example, a "5" or "7" written on a piece of paper). He also had trouble writing the numbers when the doctor dictated them out loud. Henry had no trouble reading the alphabet; it was only when numbers were involved that he seemed to be at a loss.

A CT scan revealed that Henry had suffered a small stroke in the back section of his left parietal lobe, a region of the brain that plays an important role in number understanding; it also has connections with motor areas of the brain that help us coordinate the movements of our hands, like interlocking our thumb and index finger into an "O."[31] Henry's multipurpose command center for finger movement and number understanding was down, resulting in problems for both.

Interestingly, the relation between fingers and numbers goes way beyond a shared bit of neural tissue. How we understand numbers in the first place is tied to our fingers, likely because we use our fingers when learning to count. When people are asked to indicate that they saw a number on a computer screen by pressing a keyboard key with one finger, they are better at doing this task when the finger they use matches their personal experiences of finger counting. Many people learn to count from one to five with their right hand, starting thumb first, and then six to ten with their left, again thumb first. For these right-handed one-to-fivers, recognizing a digit less than five is easier when they are asked to use their right hand on the keyboard than if they have to use their left hand. The opposite is true for larger numbers. How we count on our fingers as kids influences the way our brain processes numbers as adults.[32]

Using our fingers to count seems to help forge common ground for fingers and numbers. Children physically come to understand numbers through their fingers. The sequence of finger movements performed while counting helps children understand that each number in a sequence has a unique immediate successor and a unique immediate predecessor, except for the first. Beyond simple counting, we also use our fingers to keep track during addition, to point to objects when we are counting, and to represent cardinality (how many of something there are). A child may raise four fingers to show how old she is. The development of numerical skills goes along with use of the fingers.

According to the psychologist Brian Butterworth, a world-renowned expert in math learning, "Without the ability to attach number representation to the neural representation of fingers and hands . . . the numbers themselves will never have a normal representation in the brain."[33] Indeed children with poor fine motor skills of the fingers are more likely to experience difficulties in math later on. Finger gnosis, the ability to tell which finger someone just touched when your eyes are closed, at five years of age is a good predictor of math achievement several years later, in elementary school. It can even be a stronger predictor of math achievement than general measures of intelligence.[34] The more finger dexterity a child has in kindergarten, the better her math skills will be down the line. The opposite is also true: poor finger control is often associated with dyscalculia, difficulty in understanding numbers and how to manipulate them.[35]

Because there is a strong link between fingers and numbers, developing better finger dexterity through musical experience can improve math skills. Just learning how to use each finger to press different piano keys can help. Children who have better finger dexterity can use their fingers more efficiently to count, calculate, and show numbers of things. As a result, their math skills are often enhanced.[36]

At one time or another, most parents wonder how their children measure up to other children their age. The comparisons usually start with early motor milestones. Shouldn't my child be able to pick up her bottle by now? Sit up? Walk? It's easy to spot parents, even the ones who claim to be laid back, surreptitiously sizing up their tots against the other babies in the sandbox. These comparisons only increase once school rolls around.

Understanding how the body interfaces with the mind provides a new window into how the mind develops. Musical training can be good for developing math skills and learning to print letters helps rev up brain systems important for reading. When a child's motor system isn't able to mirror the actions she sees others performing or even contemplate the actions needed to write the letter "A" or grab a toy, it's hard for her to understand what is going on. Recognizing that children will have problems understanding what they can't do helps us see how important motor experience is for all kids—not just for meeting those all-important motor milestones but for meeting the cognitive milestones too.

CHAPTER 3

Learn by Doing

Although it is hard to tell by looking at their spongy body, sea squirts are members of the phylum chordata, which includes animals with spinal cords, such as fish, birds, reptiles, and humans. But unlike other animals in their phylum, sea squirts don't keep their brain and spinal cord forever. They keep them only for as long as they need them.

The sea squirt starts off its life cycle as a tadpole-like creature, complete with a spinal cord connected to a simple eye and a tail for swimming. It also has a primitive brain that helps it locomote through the water. Its mobility, however, doesn't last long. Once the sea squirt finds a suitable place to attach itself, whether the hull of a boat, an underwater rock, or the ocean floor, it never moves again. As soon as sea squirts stop moving, their brain is absorbed by their body. Being permanently attached to a home makes the sea squirt's spinal cord and the neurons that control locomotion superfluous, so why keep them? A brain is an energetically expensive organ to maintain, even for a sea squirt. So once the sea squirt becomes stationary, it literally eats its own brain.

Although many psychologists are comfortable with the idea

that the main function of the human brain is for thinking and feeling, the life of the sea squirt provides clues as to what brains originally evolved to do: orchestrate and express active movement. Daniel Wolpert, an engineering professor at Oxford University and winner of the famous Golden Brain Award, said in a recent TED Talk, "We have a brain for one reason and one reason only and that's to produce adaptable and complex movements. There is no other reason to have a brain."[1] And there is growing recognition that our actions and our thinking are a lot more interconnected than previously thought. Brain areas responsible for the ancient functions of navigating our surroundings and those responsible for the most novel functions, such as reading and counting, don't operate independently from one another but have plenty of opportunities to communicate and influence one another. Often these functions are rooted in the very same bits of neural tissue.

It has always been in vogue to compare the mind to the most complicated device of the day. A hundred years ago it was the telephone switchboard, used to manually connect telephone lines. In the switchboard analogy, infants' neural phone network was limited, with only a few connections in place, which explained why they knew and could do so little. As children grew, the number of their connections increased so that they thought and acted in complicated ways, their minds making a more diverse set of calls.

These days, however, the human mind is most frequently compared to a computer that has three or so pounds of neural hardware on which each of us runs many different software programs. The problem with this analogy is that, just as most software can run on any platform, seeing the mind as a computer determining our connections and interactions makes our body and physical experiences inconsequential, like tech support. Thinking is reduced to a programming language, the manipulation of symbols by rules that are carried out by hardware, not influenced by it.

Probably more than any other institution, Western mainstream education embraces the computer metaphor of the mind. Even though the information we take in comes from five different senses—visual, aural, smell, taste, and touch—educators tend to characterize the storage of this information as abstract, removed from the very senses that helped load the mind's hard drive in the first place. Lesson plans seem to be designed with the adult sea squirt in mind, as if the body is unnecessary, with students permanently affixed to their desks. Physical objects such as blocks, which help teach children about math concepts, are scarce, and even fewer objects are used to help teach reading. Students are becoming more confined than ever to their chairs.

This stationary model of education is counterproductive, because we tend to learn through movement and engaging with people and things in our environment. Take language as an example. Babies and toddlers are first exposed to language in a highly interactive context. A mom might hold up a cell phone, hand it to her toddler, point, and say "phone," or she might say the word "bottle" as she gives her child a bottle to hold. Most of the words that kids learn are tied directly to the objects the words refer to, and, more often than not, the kids get to hold and manipulate the objects they are learning about. But in standard classroom reading lessons, teachers aren't connecting what kids are reading to the physical world. Even when using picture books, many teachers focus so closely on what the words sound like that they seldom point to the pictures that depict the objects the words refer to. Reading is taught in a stripped-down fashion, devoid of a dynamic, interactive context that is integral to learning language.

Why is it a problem if words are learned without direct ties to action? One reason is that this doesn't seem to be how our brains work. Modern neuroscience has yet to find anything like an abstract, completely isolated reading area in the brain. Rather, when we read, we tend to activate the same sensory and motor brain areas involved in doing what we are reading about. When people make small body movements in the fMRI scanner, moving their feet, fingers, or tongue, they activate regions in the motor cortex involved in moving these body parts. Most interesting, when people read action words associated with the leg, arm, and mouth (such as *kick*, *pick*, and *lick*), they also activate some of these same motor brain areas. Both moving your foot and understanding the word *kick* are governed, in part, by an area of the motor cortex that controls the leg.[2] It's hard to separate the reading mind from the doing mind. Teaching words estranged from the objects and actions they refer to doesn't reflect how the brain is organized. Because our body and mind are tightly connected, the body is an important part of the learning process.

Art Glenberg has devoted his career to understanding the mental mechanics of learning. He has a full head of silver hair and a tanned face that discloses his love of sunshine and the outdoors. Glenberg retired a few years ago from the faculty at the University of Wisconsin and couldn't think of anything he wanted to do more than continue his work, so he accepted a new job at Arizona State University. Same gig, better weather—not bad. At ASU, Glenberg runs the Laboratory for Embodied Cognition. The quote on his lab's website is "Ago Ergo Cogito"—"I act, therefore I think." This motto captures how Glenberg wants to develop young readers: by incorporating movement into reading lessons to help enhance reading skills.

Because language learning involves a lot of activity, it is obvious to Glenberg that interactive reading lessons could improve kids' comprehension abilities. Just as when a dad says "Wave bye-bye" and physically waves to his kid, the children in Glenberg's studies learn to directly relate the words they are reading to the actions, objects, and events the words refer to.

In a recent experiment,[3] Glenberg recruited first and second graders to work in different reading groups. Here is a story that they worked on:

BREAKFAST ON THE FARM

Ben needs to feed the animals.

He pushes the hay down the hole. (Green light) [There is a hole in the floor of the hayloft above the goat's pen.]

The goat eats the hay. (Green light)

Ben gets eggs from the chicken. (Green light)

He puts the eggs in the cart. (Green light)

He gives the pumpkins to the pig. (Green light)

All the animals are happy now.

Some children were assigned to an "action" reading group. These kids took turns reading each sentence aloud; when they saw a green light at the end of a sentence, this was their signal to act out the events in the sentence using toy objects that had been set in front of them (a toy barn, chickens, pigs, pumpkins, a boy figurine, a cart). Other kids were assigned to the "repeat" reading group. These children also took turns reading the sentences out loud, but when they got to a green light, they simply reread the sentence.

Children who acted out the story had a better understanding of the material than the kids who simply read the sentences a second time. And these were not small differences. Acting out the sentences boosted children's understanding of a story by 50 percent or more. Those children also tended to remember a lot more details—even several days after the initial reading experience.

Of course, it's possible that acting out the scenarios simply helps engage students in the lesson, but Glenberg doesn't think so. If it is simply about paying attention, then you would expect the "repeat" group to excel. Having the opportunity to read the sentences twice should, if anything, help these kids comprehend what is going on and remember more of the story details. The explanation that Glenberg favors instead is that experience acting out the sentences pushed kids' brains to mimic those of more experienced readers. Just as when we read the word *kick* and the foot area of the motor cortex comes alive, acting out a sentence helps us connect words and their referents. Children can link what they are reading explicitly to the actions and events the words describe. When the kids are later tested for comprehension, they are able to call upon a rich set of sensory and motor experiences related to what they read, experiences that guide their memory and understanding.

Acting out their reading lessons allows children to connect words to the world around them. Kids often struggle to learn what words mean when all they get is a definition that describes that word in terms of yet other words. Glenberg's reading intervention mimics language learning in the real world, by helping kids to link words to the sorts of actions, images, or dialogues the words relate to. This action experience also allows kids to understand the varied meanings of the same word. Consider the following sentences, which conjure up different ideas about what the word "coffee" is referring to (that is, a cup of coffee versus coffee beans):

> The coffee spilled. Go get a mop.
>
> The coffee spilled. Go get a broom.

Words are more than their definition; they are defined by the context in which they occur. Actions help give words meaning, and they also help to illustrate how words vary from situation to situation. Interactive learning doesn't just offer "words for words."[4]

The importance of using the body as a tool to enhance understanding extends to other subjects besides reading. The cognitive scientists George Lakoff and Rafael Núñez have been arguing for years that children's understanding of mathematical concepts such as "add" and "subtract" develops by extending words and their related actions to mathematical situations. In fact, these scientists argue, much of mathematics, from discrete math to combinatorics, comes from the evolutionary history of the human body. We are animals with limbs that allow us to manipulate objects. Our understanding of math would be very different, they argue, if we were built like snakes, without the ability to easily hold a diverse set of things.[5]

Consider the word *add*. In physical terms, it means to place something into a container, group, or substance: "Add cream to the coffee" or "Add logs to the fire." Conversely, *take* means to remove: "Take some books out of the box" or "Take some logs off the fire." Kids learn by experience that there is a tight connection between adding objects to a collection and addition and taking objects away and subtraction. When the verbs *add* and *take* are then used in an arithmetic context—"If you add 4 apples to 5 apples, how many do you have?" or "If you take 2 apples from 5 apples, how many do you have left?"—children are able to call upon their previous motor experience to understand the mathematical concept at play.[6]

The extension of action to math helps explain another recent study by Art Glenberg, in which he found that kids who solved math story problems by acting them out were better at understanding the mathematical operations involved.[7] Consider this math problem Glenberg gave to third graders:

> There are 2 hippos and 2 alligators at the zoo.
>
> They live by each other, so Pete the zookeeper feeds them at the same time.
>
> It is time for Pete to feed the hippos and the alligators.
>
> Pete gives each hippo 7 fish. (Green light)
>
> Then he gives each alligator 4 fish. (Green light)
>
> The hippos and alligators are happy now that they can eat.
>
> How many fish do both the hippos and the alligators have altogether before they eat any?

Students who acted out the problem, who actually counted out the appropriate number of little toy fish and distributed them to the animals, were two times more likely to solve the problem correctly than the kids who simply reread the story.

But here's where the data get really interesting: a third group of students, who counted out Lego pieces whenever there was a green light, didn't do any better at solving the math problem than the kids who simply reread the story. One of the surprising lessons of this research is that it's not just any movement that produces understanding. The third graders in the Lego group were still moving objects, but these objects were unrelated to the plot of the story problem: the Lego pieces were not shaped like fish, nor were there figures of hippos and alligators to distribute the fish to. When there isn't a direct connection between words and objects, the power of action is lost.

Interestingly, the use of blocks and other objects, or manipulatives, is becoming more and more popular in classrooms across the nation (especially in more elite schools): students are taught to count with blocks or sticks as a way to solve math problems. Originally created in the early 1900s for educational use, block play is being touted by teachers and parents alike as the new cure-all for our educational woes, and national school suppliers have added a ton of new block-related products to their catalogues in the past several years. Private schools now use their blocks as a recruiting tool.[8] Manipulatives are even advocated by the National Council of Teachers of Mathematics as a way to enhance students' grasp of basic math concepts like subtraction and addition.[9] Yet while the block movement represents a renewed faith in incorporating active play into learning, how exactly this block play is carried out determines what kids learn. It's not simply the handling of blocks—or Legos, as we saw in Glenberg's study—that's important. Rather, as Glenberg's work clearly shows, manipulatives have a positive learning benefit when they can be directly connected to the content of the problem students are trying to solve.

Why does the direct linking of children's actions to the story content matter? Consider the word *each,* which Glenberg thinks children have a particularly hard time with. Understanding this word is actually quite complicated: the word must be connected to the correct set of objects, and the objects within the set need to be seen as distinct entities. It is not enough when reading *each* to note that there is a group of alligators. The reader must also realize that there are *two* alligators and that they are fed individually. Physically manipulating the relation between the fish and the characters in the story makes this individuation pretty clear, because the child has to count out fish for each of the alligators. It's less obvious when kids don't do this sort of story-relevant counting. In fact Glenberg found that the most typical error among kids who counted with Legos was to say that the hippos and alligators had eleven instead of twenty-two fish before they ate any of them. It's as if the kids failed to realize that *each* meant that the eleven fish had to be doubled to get the total for the two alligators and two hippos. By acting out the story with relevant manipulatives, children come to understand symbols (such as the word *each*).

Random hands-on activities are no panacea for educational woes, but carefully structured action experiences can help children learn. Kids don't have to walk around with a toolbox of toys for math and reading in order to get an action benefit. Glenberg and his research team have also shown that, once children have some action experience, they can imagine performing the actions in the stories and still get a benefit. When the connections from words to actions are in place, it is easy to capitalize on them.

Of course, cognitive scientists weren't the first to tout the educational benefits of movement. Maria Montessori, the founder of the Montessori educational movement, wrote a hundred years ago, "One of the greatest mistakes of our day is to think of movement by itself, as something apart from the higher functions. . . . Mental development must be connected with movement and be dependent on it. . . . Watching a child makes it obvious that the development of his mind comes about through his movements. . . . Mind and movement are parts of the same entity."[10]

In Montessori schools, kids learn the alphabet by tracing letters and, just as in Glenberg's reading lessons, learn grammar and vocabulary by acting out sentences their teachers read to them. For decades the emphasis that the Montessori method placed on a dynamic learning environment was largely ignored by mainstream educators, but recent advances in neuroscience and psychology show how critical movement is for understanding. This new research in embodied learning helps provide a road map for how to structure educational activities to best help kids learn. The mind is not an abstract information processor largely divorced from the body and the environment. It is highly influenced by the body and movement.

In a math class called Math Dance, people move in a circle around the room to the rhythm of the beat, while a leader sits in the middle playing bongos. Developed by the choreographer Erik Stern and the mathematician Karl Schaffer, Math Dance is a series of whole-body mathematical activities.[11] "Many math-phobic adults and children—young people—are put off by math because they are given symbols before they have a real solid experience on which to base it," Stern explains.[12] Math Dance is designed to give people the physical experience of an abstract idea. By translating math into movement, students and their teachers may be able to better understand numbers.

Schaffer and Stern met over twenty-five years ago in Santa Cruz, California, through dance. At the time, Stern was dancing with the troupe Tandy Beal & Company, a popular group in the northern California performing arts scene. Schaffer was working on a PhD in mathematics at the University of California, Santa Cruz, but he also spent a lot of time in the Dance Department.

The two hit it off and several years later began creating works that explored the connections between math and dance.[13]

In 1990 they put on their first math dance production, called *Dr. Schaffer and Mr. Stern: Two Guys Dancing about Math*. The performance was so popular that they soon began touring the country, putting on their math dance for schools and educational organizations. Before long, teachers started asking if some of the activities in the performance could be used in the classroom. So Schaffer and Stern set out to translate their act into a series of in-class math activities, which became Math Dance.

They started with an activity called Counting Handshakes, taken straight from the opening dance in their performance. As Stern and Schaffer put it, their opening is an almost "vaudevillian" handshake sequence, in which the two characters can't seem to find a way to shake hands; when they finally figure out how to come together for the handshake, they realize they are stuck. When the dancers first started doing the performance, they were struck by the different ways two people could actually shake hands. The Counting Handshakes activity explores the mathematical concept of combinations. In pairs, students create a movement sequence by discovering the variety of handshakes two people can do using one hand at a time. For instance, the first person might start by using her right hand to shake the second person's left hand, then left to right, left to left, and right to right. One obvious answer, given that each student has two hands, is that there are four different possible combinations. But students get creative, using secret handshakes to increase the number. In this way, students learn the concept of discrete entities.

Discrete entities, such as handshakes or dogs, occur only in whole units, unlike water or the height of a tree, which can be measured in fractions. Although students may not realize it at first, by engaging in the simple Math Dance handshake, they are doing discrete math, called combinatorics, the area of mathematics that deals with counting combinations of things. Experiencing the physical element helps students understand the abstractions of math, specifically what it means for something to be a distinct entity.

Understanding combinations of things and how to test all the possible permutations helps students grapple with difficult math concepts they will encounter from elementary school through college. Take the following algebra problems a middle schooler might see:

> John has two shirts and three pairs of pants. How many possible outfits does he have?
>
> Answer: 2 × 3 = 6 possible outfits (as long as John isn't a nudist and always wears a shirt and pants)
>
> Sally has a 6-CD player in her car and 100 CDs. How many unique ways can she load the player?
>
> Answer: There are 100 ways to choose the first CD, 99 ways to choose the second, 98 ways to choose the third, 97 ways to choose the fourth, 96 ways to choose the fifth, and 95 ways to choose the sixth. So 100 × 99 × 98 × 97 × 96 × 95 = 858,277,728,000 (as long as Sally always loads 6 CDs at a time)

Students who are able to physically experience the concept of *discrete* are better equipped to link these equations to their context, even drawing out the different possible combinations as a way to determine whether the algebra equation they derived is correct. Just like Glenberg's third graders counting out the fish to give to *each* of the animals in the math story problem, understanding the concept of *discrete*, and that a finite number of possible combinations exists, helps ground the meaning of abstract algebra in something concrete.

In another Math Dance exercise, each student starts by creating a movement. Then students pair off and toss a coin 10 times. Heads represents one partner's movement, tails the other partner's. Before they actually toss the coin, students predict how many times they will each have to enact their movement. Before the exercise, most students assume that they'll be doing each movement roughly an equal number of times. But they soon discover this isn't the case and that the 50 percent probability of hitting heads versus tails doesn't actually come true, at least until you have done several thousand iterations. Kids learn that getting close to 50 percent gets more likely the more tosses they do—a key component of the concept of probability.

Perhaps most surprising about Math Dance is that the movement itself matters a lot. Doing the dance along with the coin tosses is an important part of Schaffer and Stern's lesson on probability because, when we move, we often remember concepts and ideas better than when we stand still.

Dancers have long recognized the power of the body as an aid to memory. When ballet dancers learn new choreography, they physically act out the movement sequences to commit the steps to memory. When asked to recall what they learned, dancers tend to recall the dance movements in chunks, based on which body positions flow together. They use their body as a mnemonic device to help them organize their steps, which makes the steps easier to recall. In the same way, performing movements tied to math concepts helps students "choreograph" the math, helping them understand how different concepts fit together, which makes each concept easier to load into memory.

Physical performers besides dancers understand the link between body and mind. Athletes from figure skaters, to gymnasts, to Olympic-caliber divers know every inch of their body and also know that the amazing tricks they perform are based on principles of math and physics. Take the British diver Tom Daley. After wowing the international diving scene at the 2010 Commonwealth Games in Delhi with two Gold Medal performances, as well as his boyish good looks and charm, he was expected to repeat his wins at the 2012 London Olympics. The problem was that Tom was only sixteen and still growing. "I'm five foot eight and if you get taller than five foot 10 then there could be some problems," he told a BBC interviewer after his winning performance in India. "If you're too tall you start spinning too slowly so you can't fit in all the rotations in time before you hit the water. You just have to cross your fingers and hope you don't grow too tall."[14]

By the time the 2012 Olympic Games rolled around, Tom had grown over an inch, to 5 foot 9½. Thankfully, with a spectacular last dive that ensured his place on the podium, Tom walked away with a Bronze Medal in London and the love of his hometown crowd. David Beckham texted him congratulations, and Prime Minister David Cameron went to see him.[15] But it wasn't an easy road to victory. Tom had to learn several new dives in the years leading up to London to ensure that, despite his growing frame, he would be able to do multiple rotations so that his dives would be graded at a high level of difficulty. There is no doubt that his coaches' and his own understanding of physics was crucial in coming up with his new repertoire of dives.

A grasp of physics helps athletes understand how best to

move and rotate their body, but how we move can also aid how we think about math and science in the first place.

Susan Fischer bounces around at the front of her introductory physics class at DePaul University in Chicago, desperately trying to get her students interested in the topic of the day: moment of inertia. But she's not having much success. It is fall in Chicago, which means that lots and lots of ice and snow are just around the corner, and Chicagoans make it a point to cherish every last sunny day. Students are oscillating between attending to the lecture and looking out the two large picture windows in the left-hand wall of the lecture hall, where the warm sunlight is streaming in. From my vantage point in the back row, I can also see that there is a fair amount of email checking and Internet surfing going on. The girl sitting directly in front of me is even buying a pair of shoes from Zappos.com. Until, that is, Fischer puts up a pop quiz on her PowerPoint. All of a sudden, everyone looks up—panicked. Even the shoe buying stops.

Here's the question Fischer put on the screen:

> A solid Disc and a Ring, both of equal mass and diameter, are held at the top of a wooden ramp. When released, both will begin to roll down the ramp without slipping under the influence of gravitational force. If the Disc and Ring are released at exactly the same time, which of the following statements are true?
>
> A. The Disc will arrive at the bottom of the ramp first.
> B. The Ring will arrive at the bottom of the ramp first.
> C. The Disc and the Ring will arrive at precisely the same time.

Except for the sound of students rustling through their bags and purses to find their clickers (hand-held devices that allow instructors to test students on the fly), there is total silence in the room. But when Fischer announces that the clickers won't be needed, there is an audible and collective sigh of relief. She tells her students that they are going to use their body to figure out the answer. Teaching assistants appear in the aisles, handing out plastic rulers and black binder clips to each student. I am given a ruler and a binder clip too. Fischer tells us to hold the ruler at one end between our thumb and index finger and to feel how easy it is to make the ruler bounce up and down. She then directs us to attach the binder clip to the opposite end of the ruler. "Now do the same thing," she says. All of a sudden it becomes much harder to make the ruler bounce. Fischer then demonstrates that as you move the binder clip closer to where your thumb and index finger are positioned, the ruler becomes easier and easier to bounce up and down. You can actually feel the difference, and, when students are eventually told to vote for their answer to the pop quiz question, an overwhelming majority of the class gets the problem right. (The answer is A, by the way.)

Fischer says students never understood the Disc and Ring problem before she added the interactive element. That's why high school physics classes sometimes go on field trips to amusement parks: experiencing moment of inertia firsthand while riding upside down on a roller coaster, screaming, gives physical meaning to an otherwise abstract concept.

Like mass, moment of inertia is a property of an object. But unlike mass, which an object has regardless of how you look at it or manipulate it, moment of inertia depends on how far the mass of the object is distributed away from its axis or point of rotation. The closer the mass is to the point of rotation, the smaller the moment of inertia, and the easier it is to move. That's why the ruler with the binder clip attached is easier to bounce up and down when most of the mass (the binder clip) is close to the axis of rotation (in this case, a student's thumb and forefinger). It's also why the Ring will arrive at the bottom of the wooden ramp after the Disc. As long as the Disc and Ring are of equal mass, the Ring must have a greater moment of inertia, in effect, making it harder to get rolling, so it arrives at the bottom after the Disc.

Fischer thinks that teaching students to feel the properties of moment of inertia firsthand helps to engage the motor areas of their brain that are used to registering mass and rotation in daily life. After all, our motor system evolved to help us deal with rotating objects and to aid us in manipulating tools of different masses. Getting our brain's action centers involved in contemplating physics concepts based on movement is the best way to learn.

Fischer didn't stumble accidentally on the power of experiencing. You wouldn't know it from her fairly tall frame, but she was once a competitive diver—a sport, like gymnastics, in which being taller is a disadvantage, as Olympian Tom Daley noted. In diving, the degree of difficulty and thus high scores depend on doing lots of rotations in the air. The taller you are, the greater your moment of inertia, so the slower you spin and the fewer rotations you can do. In other words, if you are too tall, you spin too slowly and can't get in all the rotations you need before you hit the water. That's why figure skaters spin so quickly when they are all tucked in. When you bring your hands in toward your body, you are shrinking the moment of inertia, so you spin faster; when the hands come up and out, you slow down. In diving, Fischer was the Ring and all her smaller competitors were the Disc.

Fischer has her students become figure skaters themselves by sitting in a spinning chair with their feet off the ground. If you hold a book in each hand and extend your arms, then pull your arms in toward your body, the spinning chair speeds up. She is convinced that, when students feel this change in moment of inertia, when they engage their body in understanding the concept, they do better on tests of this concept. She told me about this idea a few years ago, when we were first introduced at a gathering of women scientists in Chicago. I was intrigued by her theory and the idea that our bodily experiences can affect our thinking, so I offered to help her test her hunches (with the help of one of my graduate students, Carly Kontra).

We have found that becoming part of a physical system enhances learning. We've had students move their arms while on a spinning chair, engage in the binder clip experiment, and move an axle with a bicycle wheel spinning on it from vertical to horizontal and back, thus changing the direction that the wheel is spinning in the world. Indeed compared to just watching a demonstration in class or simply reading about the physics of mechanics in a textbook, bodily experience leads to marked learning gains on homework assignments, quizzes, and tests that can be seen even weeks later.[16]

Why? Using fMRI to peer inside the brains of students who have actively engaged with physics concepts like moment of inertia, angular momentum, and torque, Fischer and my research group have found that the motor cortex, the chunk of brain tissue involved in planning and initiating movement, is activated. After physically experiencing these concepts, students later activate their motor cortex when they just think about, say, angular momentum when taking a quiz on the topic. It's as if their motor system is replaying their previous experiences, helping them reason about what they can't actually see and feel in the moment. The more the motor cortex is involved, the better students do on related test questions about the physics of mechanics.

In short, getting the body involved helps the mind learn.

CHAPTER 4

Don't Just Stand There

HOW MOVING SPARKS CREATIVITY

Moved to Insight

Google's corporate headquarters (otherwise known as the Googleplex) is set on twenty-six acres in Mountain View, California. There are four main buildings, each of which houses an eclectic mix of computer scientists, engineers, and managers. While it might seem most straightforward to group Google employees by what they do—engineers in one building and managers in another—that isn't how the company works. Instead the space is designed to foster an atmosphere of interaction. People with different jobs are mixed together across the Google campus, and structures such as an indoor tree house and a volleyball court encourage employees to get up and move.

Google says the whole point of their campus's design is to encourage interactions between different teams of Googlers, to spark conversations that might not normally take place. But the interactive atmosphere also encourages movement. We have already seen how movement can help kids learn and adults remember. Movement also helps us solve problems and can even increase our productivity because thinking involves moving the body as well as the mind.

To get a better sense of how movement can enhance problem solving, consider the following scenario:

> You are a doctor and have realized that your patient has an inoperable stomach tumor. There are certain lasers that can destroy this tumor if their intensity is great enough. That's the good news. The bad news is that, at the intensity the lasers need to be to destroy the tumor, they will also destroy the healthy tissue that surrounds the tumor. The tumor is malignant, so if you don't operate on it, the patient is going to die. How can you destroy the tumor without damaging the healthy tissue through which the lasers must travel? Is there a type of procedure you can do that will obliterate the tumor while at the same time making sure that the healthy tissue that lives around the tumor isn't damaged?

If you conclude that your patient is screwed, you are in good company. This problem proves to be a difficult one to crack. In fact only around 10 percent of undergraduate students get this problem correct when asked to solve it the first time around.[1] But there is a pretty simple way to increase success, and—as you may have guessed—it involves the body. People who are given a computer diagram of the problem (a circle depicting the tumor inside surrounded by a thick layer of healthy tissue outside) and are asked to contemplate possible solutions, but who are also asked to simultaneously keep track of a tiny dot bouncing around the screen, are much more likely to come up with the right answer—as long as the tiny dot is moving back and forth through the healthy tissue to the tumor at various points around the tissue's perimeter.[2]

In case you haven't figured it out yet, the solution to the problem is to position a number of separate lasers around the patient, each directed at the stomach tumor. If each laser delivers a small amount of the radiation, you end up with enough accumulated radiation to destroy the tumor while at the same time saving the healthy tissue surrounding it.

By moving the body (in this case the eyes) in a way that mimics the solution, people have thoughts about the problem that they wouldn't have otherwise had. Students think the dot task is designed to distract them, that it will make it harder to solve the tumor problem. But when the eyes draw a path that shows many different lasers converging on the tumor from many different areas, the bouncing dot actually leads to increased success.

Moving the body can alter the mind by unconsciously putting ideas in our head before we are able to consciously contemplate them on our own. People use their body all the time when problem solving, without even knowing it. In the case of the tumor problem, researchers found that we often unconsciously work our way through the scenario, testing out possible solutions with our eye movements. Most interesting, before students realize they have come up with the correct answer, you can actually see the solution in their eye movements.[3]

What accounts for this direct link between body and mind? We draw on our concrete physical experiences to construct our reality. For example, the tactile sensation of warmth causes us to think about social closeness, and making a fist leads us to feel more assertive. Getting a person to move lowers his threshold for experiencing thoughts that share something in common with the movement. That's why moving your eyes in a way that mimics the solution to the tumor problem heightens the probability that you will find that solution.

Sometimes the best way to crack a problem is to get moving. This is advice that dancers have followed for years. They constantly use movement to create new ideas. When dancers try to develop a new movement, their body is their medium, similar to the way an artist might use paint or a violinist uses the sounds of her violin to create. Just as choosing a different instrument or shifting from paint to pencil changes the art form, so too does changing the body. Make a body rigid, and the style and form of the dance are altered. The mechanics of the body are front and center in creativity. The thinking process is extended over the body. In other words, many performers literally think with their body.[4]

More evidence that our actions influence our thoughts—and specifically our ability to be creative—comes from research on metaphor. We use metaphors constantly, whether it is thinking "outside the box" or "putting two and two together" or first considering a problem "on the one hand, then on the other."

But here's where things get really interesting: our creative ability is enhanced when we *literally* act out creative metaphors. When people are asked, say, to come up with a word to add to *measure, worm,* and *video* to form three new compound words, most find the task difficult. (The answer is *tape: tape measure, tapeworm,* and *videotape.*) Finding the answer involves searching our mind for some broad connection among the words, a creative way to fit them together. It's not easy. However, when people literally embody the "thinking outside the box" metaphor, they become more creative and better able to solve these types of compound-word puzzles.

To demonstrate the physical reality of creative metaphors,

researchers at Cornell University constructed a box out of PVC pipe and cardboard that measured five feet on each side. Thus the box could comfortably contain an individual seated inside. The researchers placed the box in their laboratory and asked volunteers to complete ten compound-word puzzles like the one above while sitting either inside or outside the box. To make it seem reasonable that people were being asked to sit in a box, the volunteers were told that the scientists were studying how different work environments affect thinking. Amazingly, people sitting outside the box solved more of the word puzzles than those sitting inside the box or when there was no box at all.

The ten word puzzles that the researchers used were:

1. Print-Berry-Bird ____________
2. High-District-House ____________
3. Fish-Mine-Rush ____________
4. Basket-Eight-Snow ____________
5. Mouse-Bear-Sand ____________
6. Cadet-Capsule-Ship ____________
7. Fur-Rack-Tail ____________
8. Hound-Pressure-Shot ____________
9. Flake-Mobile-Cone ____________
10. Safety-Cushion-Point ____________

Answers:

1. Blue
2. School
3. Gold
4. Ball
5. Trap

6. Space
7. Coat
8. Blood
9. Snow
10. Pin

People were also more likely to come up with clever captions for pictures and to generate more unique ideas about unfamiliar objects when walking freely than when sitting down or walking in a box formation.[5] So next time you are trying to come up with the clever caption for the cartoon that will be printed in the *New Yorker*, get up and take a walk. It might be just the movement you need to bring that amazing caption to mind.

When we read something confusing or have to find a solution to a difficult problem, our first instinct is often to sit down, to stop whatever we are doing in order to concentrate. We rarely consider what we are doing with our body. But being sedentary may be the worst thing you can do. The literal and abstract meanings of creative metaphors have become so interwoven that these metaphors have taken on a physical reality of their own. That's why acting out creative metaphors can give rise to novel ideas. Literally thinking outside or without physical constraints (walking outdoors, pacing around) may help facilitate new connections between distant ideas, which is what creativity is all about. Indeed my colleagues and I have joked that one of the best things about becoming a faculty member is not having our own office or a (somewhat) bigger paycheck than we got when we were graduate students, but the fact that during class we no longer are confined to our seat at the seminar table. We can walk around as we think; we can use the fluid movements of our body to help free our mind from constraints.

We look to our physical experiences to create reality. Perhaps that's why Chinese exercise balls (otherwise known as Baoding balls) became a staple on executives' desks. Most people think of them as a stress reliever or just something to do with their hands when they are on the phone or in a meeting, but those small shiny silver balls that people move from hand to hand likely serve a much bigger function: improving our creative thinking. Physically moving those Baoding balls from one hand to the other may help us think about an idea on "one hand and then another." Dynamically coordinating our hand movements can facilitate the mental mechanics of creative problem solving, helping us to see a problem from multiple perspectives. The unexpected benefit of creative thinking that comes from moving our body reveals the importance of physical actions to improve performance at work. We live in an age when it is easy to be static, at our desk, on the elevator, or in a meeting, but being motionless can inhibit our thinking.

Our actions also affect our ability to take charge and get what we want. Being able to achieve at work, to be productive, is really about sending your brain the signal "I am in charge. I feel good, go," says Dana Carney, a professor at the Haas School of Business at the University of California, Berkeley.[6] One way to send that message is to adjust the body. Carney and her colleagues Amy Cuddy and Andy Yap have found that, when people hold open, expansive postures, otherwise known as power poses, they put themselves in a better state of mind. These power poses may also increase the level of testosterone circulating in the brain and body. Testosterone, a sex hormone, is often a culprit in doping scandals in sports, when athletes pump their body full of testosterone as a quick way to improve muscle mass and strength. But this hormone affects the brain too. Increases in testosterone have been linked to increases in confidence, attention, and memory. Testosterone is also involved in competitiveness and risk taking, giving you the confidence to approach and solve problems. When you have to go out on a ledge to offer a novel solution to a problem, something as simple as how you hold your body can help convince you and those around you that your viewpoint has merit. Carney and her colleagues have found that a one-minute power pose leads to roughly the same amount of testosterone increase that most people experience when they win a game.[7] In short, power poses can make the difference between holding your own in a meeting and coming out on top or succumbing to a bad deal.

You can tell a lot about how a person is feeling from how she holds her body. People who feel anxious tend to move their body in less natural ways and lean away from their interaction partners, even when they swear they are not feeling anxious. It's hard to lie about how we are feeling when our body gives us away.[8]

Surprisingly, body cues can often be more important than facial expressions in indicating how a person is feeling. A few years ago, a group of Princeton University psychologists put together three groups of photographs of professional tennis players like Maria Sharapova and Andy Murray, taken after they had either won or lost a major point in a high-stakes match: one group contained full-body shots with faces showing, the second group contained face shots without bodies showing, and the third group contained body shots without showing faces. When the researchers asked volunteers to guess what emotions the tennis players were experiencing at the time of the photo, they discovered that viewers were much more accurate at guessing the players' emotions when they could see the players' body, with or without also seeing their face.[9] People physically communicate what they are feeling not just with their facial expressions but with their entire body. Perhaps that's why we tend to be so interested in the postures people display in situations—especially competitive ones—ranging from sports to the business world.

After the Jamaican sprinting legend Usain Bolt set a new Olympic record in the men's 100-meter sprint in the 2012 Summer Olympics, he stretched one arm to the sky as if he were about to send a lightning bolt upward. This open body posture was immediately copied around the world and sent a strong signal about who was in charge on the track. Mario Balotelli, a soccer player for the Italian National Team, struck a pose after he scored a goal at the 2012 European Championship. Picture a mohawked, bare-chested striker, with a wide stance and his arms held out wide and flexed downward at his side, fists clenched, and a look on his face that says, "I can score at will." His pose reflected who was in charge and also likely helped him feel at the top of his game. It's easy to find photos on the web of adults, kids, and even pets mimicking former NFL quarterback Tim Tebow's signature pose: bending down on one knee as if you are about to propose to the love of your life and then, with an elbow resting on your bent knee, flexing your arm and clinching your hand into a fist. Tebow may have been thanking God for a good play, but his open stance also sent a powerful message to his brain about his ability to dominate (albeit, short-lived) on the field. Finally, consider the wide-legged stance that Donald Trump often holds when he is talking to contestants on his TV show, *The Apprentice*, or merely giving a television interview about a new development in the works. Taking up a lot of space with your body sends signals about your confidence and dominance. By simply assuming a pose, you can change how you are thinking and feeling.

Expansive, open postures tend to increase our feelings of

power and control. They can even enhance our ability to project power and confidence to others. There is more to how we hold ourselves than meets the eye. Here are a few body position tips from the power posing research, whether you are in a meeting, pitching to a client, negotiating on the phone, or by yourself; they capitalize on the idea of spreading your limbs and taking up a lot of space:

If you are standing at your desk, stand tall and keep your hands wide apart.

Rest your arm on the chair next to you during a meeting. It opens up your body, allowing you to take up more space.

Put your hands behind your head with your elbows pointed outward.

Uncross your legs, even prop them up in front of you on a desk or table. This helps you take up additional space and creates a mental and physical feeling of expansiveness. Try this technique when you are on the phone as a way to feel larger physically and emotionally, more confident and assertive.

No matter what pose you choose, wear comfortable clothes that aren't too constraining. Whether you are answering email, on an important call, or in a heated meeting, you want to feel that you can stretch out and open up physically and mentally.

Before you assume your most expansive pose, a caveat: It's true that expansive body postures lead to feelings of power, which can be a good thing. However, as we know from history and politics, power can lead to dishonest behaviors—cheating, stealing, and other forms of corruption. One study found that people who routinely struck expansive poses (with legs widespread and arms out wide with their hands resting on their hips), as opposed to contracted ones (standing with their legs and arms crossed), were more likely to keep money that they were "accidentally" overpaid. People sitting at a desk in an expansive pose were also more likely to cheat on a test than those in a more contracted posture. Even those who drove cars with a more expansive driver's seat (allowing them to spread out when they were behind the wheel) were more likely to be illegally double-parked on New York City streets than drivers with smaller seats.[10]

The take-away is that expanding your body can be a powerful psychological booster. Just be careful not to wield this power for the wrong reason.

Remembering

Because there is such a close connection between physical and mental activity, moving the body can change how you think. Whether you're moving your eyes, walking freely, or even assuming different postures, your body can alter your mind and influence what others think of you. Just as dancers use their body in order to remember and communicate their role, actors recognize the value of the body for communicating ideas and for memorizing scripts. When actors learn their lines, they don't just concentrate on the words on the page but also imagine the moment-to-moment actions of the character uttering them. Actions help infuse memories with an emotional charge that makes them last. Our previous experiences being in the world affect how we understand what we see, hear, and read about.

Though actors rarely consider memorization as a defining feature of what they do, the ability to remember large amounts of dialogue and effortlessly deliver their lines on cue is a pretty amazing feat. A nonactor might imagine that learning a script involves days, weeks, or even months of rote memorization, but actors don't work that way. As one professional put it, "Most of the time I memorize by magic—that is, I don't really memorize. There is no effort involved. There seems to be no process involved: It just happens. One day early on, I know the lines."[11]

How does this work? One of the greatest mysteries of the human mind is how we remember information so that we can later recall it on cue. This is something most people struggle with, not just actors. In school, students spend a lot of time committing information to memory for important exams; lawyers grapple with memorizing an opening or closing argument so that they can deliver it smoothly and convincingly in the courtroom; and executives must commit to memory key elements of presentations to boards, customers, and employees to make sure their pitch goes off without a hitch. But actors memorize their lines and go "off book" by using the body as a tool. They try to actively experience the words of the character they are playing, which often involves connecting a specific dialogue to a specific action. Studies have shown that dialogue spoken while performing an action, like walking across a stage, is more readily remembered than are lines unaccompanied by action. Even months after a final performance, actors recalled lines accompanied by movement better than lines spoken while sitting still.[12]

Why is this active experience so important for committing information to memory? Consider the following situation: A character walks across the stage, picks up a bottle, and remarks, "This is how I solve my problems." The actor knows why the character is saying what he is saying, which affects how he walks toward the bottle and even how he grabs it. Whether the character is going to take a big gulp from the bottle, throw it at another character, or pour the contents into the sink changes his actions. The bottle represents the meaning of the situation, and the meaning of the situation constrains what actions he will perform. The opposite is also true: how the character handles the bottle constrains the meaning of the situation and the words that he might utter. It would be odd if an actor remarked that the bottle contained a prized and very expensive wine and then threw the bottle in the trash. When actors later retrieve a scenario from memory, they recall both the dialogue and their actions. Because the particular actions they performed limit the range of possible dialogue, they are better able to remember their lines. The specificity of an actor's sensory and motor experience aids recall of the literal words. Memory is grounded in the body.[13]

It's no secret that, as we get older, our ability to remember specific bits of information decreases—a major source of frustration for older folks (and those around them). But something as simple as taking an acting class that pushes us to become active experiencers in daily life can help combat memory problems. Indeed seniors who took a month-long acting class had a better memory than those who participated in an art appreciation class, precisely because the acting class taught them to be active experiencers. People took the strategies they learned in the acting class and applied them in their everyday lives, using movement as an extra hook for remembering all sorts of things.[14]

Whether you are giving the wedding toast in front of two hundred of your closest friends and family, making an important speech to colleagues, or simply recalling how many kids your boss has, use your body to aid your recall. Practice picking up a glass as you prepare your toast, incorporate hand gestures into your speech, index the number of your boss's kids on your fingers.

That way, when all eyes are on you and you have to remember your lines, your body will be in a position to do some of the remembering for you. Memory isn't confined to what goes on from the neck up; your entire body plays a powerful role in the memory process.

Here are two more memorization tips you can use, whether at school, in the office, or on the stage:

Test yourself. Practice recalling the information you are going to need for that test or pitch to a client as a way to try to learn the material. We rarely think of tests as learning events, but a wealth of studies shows that they are. Testing seems to help us load information into memory in a number of ways: it helps us relate what we already know to what we are learning so that we have many different hooks by which to retrieve information at a later time; it helps us mentally organize what we are learning; and it helps us figure out what we still need to learn and pay attention to in subsequent study sessions.[15]

Spread out your practice sessions. Whether it's for a test or for a big presentation, at one time or another most of us have left our preparation to the last minute. While cramming for a test is better than not studying at all and can help you recall some information in the short term, when you distribute your attempts to commit information to memory over time (called distributed learning), you are more likely to remember information in the long term.[16] This is because when we put more space between our learning episodes, we have to work harder to remember the information each time—which helps us better commit it to memory.[17]

Our body helps us remember information and helps us communicate with others. Hand gestures, which we focus on in the next chapter, in particular are a frequent aid in conveying information to others. We use them to give directions, signal how we are feeling, and provide added meaning and emphasis to our spoken words. Yet we gesture even when no one is watching, such as when we're on the phone. How we move our body provides a hidden window into our mind. But our body movements can also change our mind.

CHAPTER 5

Body Language

HOW OUR HANDS HELP US THINK AND COMMUNICATE

We Think with Our Hands

On the evening of October 15, 2008, Senators John McCain and Barack Obama took the stage at Hofstra University in New York for the third and final debate of the 2008 U.S. presidential elections. With less than three weeks until election day, it was the candidates' last chance to get their message out on the national stage. McCain, who was trailing Obama by roughly 8 points in the national polls,[1] started out aggressively, challenging Obama on everything from his politics to his character.

Obama had expected a heated debate; even though he was ahead in the race, the pressure was on him too. A few days earlier, on the campaign trail in Holland, Ohio, the Democratic nominee had had a run-in with Samuel Joseph Wurzelbacher, a tall, balding man with a booming voice who confronted Obama about his small business tax policy. "Joe" remarked that he was about to buy the company he worked for, a company making more than $250,000 a year. He was concerned that Obama would tax him more if he acquired the business. Media circles were instantly captivated by Joe's challenge. Overnight "Joe the plumber" became a household phrase. As it turned out, Joe wasn't even licensed as a plumber in the state of Ohio, nor did he have any immediate plans to buy the plumbing company that employed him.[2] But none of that mattered. Obama was going to have to address Joe's concerns head-on in the debate.

Approximately 29 million viewers tuned in to watch the debate that night. It was sure to be good television, and voters were keenly interested in hearing each candidate's plan to revitalize the economy. McCain wasted no time bringing up "Joe the plumber," evoking his name over a dozen times in the first half of the debate alone. Obama's plan was "going to increase [Joe's] taxes," said McCain, gesturing pointedly with his *right* hand. Obama countered, emphasizing his view by beating his *left* hand on the table in front of him: "I want to provide a tax cut for 95 percent of working Americans, 95 percent." Essentially Obama was saying that it was better to lower taxes for Americans who make less so they could afford to do more.[3]

Americans were highly attuned to what each candidate was proposing at this late stage in the game, but not everyone was simply *listening* to their speeches. The psychologist Daniel Casasanto was also observing their body language, especially their hands. People gesture with their hands all the time while speaking, often without realizing that they are doing it. Casasanto—at that time a postdoctoral fellow at Stanford University—was doing research to try to better understand why we gesture while we speak. How do a speaker's gestures help him effectively communicate a message to others? What exactly do these hand movements reveal? Casasanto had a hunch that hand gestures serve as a window into what a person is really thinking. Unlike the words we utter, our gestures tend to be more automatic, not under our conscious control. Casasanto thought that gestures might reveal what people were hesitant to put into words, and he was particularly interested in the specific hand that politicians gesture with when speaking about touchy topics like health care and tax reform.

Casasanto observed that, throughout history, there has been a tendency to associate the right side of things with "good" and the left side with "bad." When people gesture with the right hand, this tends to indicate that they have a positive view of what they are saying; the opposite is true when they use their left hand. In ancient Rome, orators were warned never to gesture with only their left hand while giving a speech; in modern Ghana, pointing and gesturing with the left hand is taboo. According to Islamic law, the left hand should be used only for dirty jobs, like wiping oneself after using the toilet, whereas the right hand should be used for eating; likewise one should enter the bathroom left foot first and enter the mosque right foot first. As in English, the word for *right* in French (*droit*) and German (*recht*) denotes both the direction and, as a noun, a legal right or privilege. This is in contrast to *gauche* and *links*, the French and German words for *left*, which are associated with words meaning distasteful or clumsy. Perhaps the author of Ecclesiastes summed it up best in this adage: "The wise man's heart is at his right hand, but the fool's heart is at his left."[4]

Why is *right* connected with good and *left* with bad? Casasanto believes that it comes from our experiences interacting in the world. Bodies are lopsided, not symmetrical, and most people have a dominant hand. Activities like signing our name or putting a key in a lock are more easily—more *fluently*—performed with our good hand. Interestingly, this fluency influences people's evaluations of objects and other people: we like things better when they appear on our dominant side. When right- and left-handers were asked to judge which of two products to buy or which of two job applicants to hire based on brief descriptions found on either the left or the right side of a page, right-handers tended to choose the person or product described on the right, but left-handers chose the person or product on the left. We tend to favor the side that we find it easier to act in. In short, the content of our mind depends on the structure of our body, and different bodies lead to different ways of thinking.[5]

This means the story is a bit more complicated than *right equals good* and *left equals bad*. Casasanto believes that the linking of *right* with good ("my right-hand man") and *left* with bad ("my two left feet") likely developed because the right-handed majority prefers to interact with the world using their right hand (and the right side of their body more generally). Of course, this also means that for the minority left-handers in the world, these associations are reversed. In other words, even though the hand with which people gesture might tell us what they think about the message they are conveying, how we interpret their gesture depends on their preferred hand. Right-handers will tend to gesture about good things with their right hand and bad things with their left. The opposite is true for lefties.

As Casasanto watched the final 2008 presidential debate, he realized he had the makings of a perfect natural experiment. He could actually test whether or not the candidates were more likely to gesture about positive messages with their dominant hand. But there was one small problem: McCain and Obama are both left-handed. Casasanto was missing a right-handed candidate to serve as a comparison. Fortunately all he had to do was turn the clock back four years to a video of John Kerry and George W. Bush, who ran for the U.S. presidency in 2004. Both are right-handed.

Casasanto ran a simple test: he analyzed both the speech and the gestures of all the candidates in the final debates of the 2004 and 2008 presidential elections to see whether they were more likely to deliver positive messages when gesturing with their preferred hand and negative messages when gesturing with their nonpreferred hand. Casasanto and his research team combed through more than three thousand spoken sentences and almost two thousand gestures. The findings were clear. For Obama and McCain, left-hand gestures were associated more strongly with positive statements and right-hand gestures with negative ones. The opposite pattern emerged for Kerry and Bush.

Democrats are said to be on the left side of the political spectrum, and Republicans on the right. Yet Casasanto found that gestures related to good and bad sentiments depended on the handedness of the politician, not his politics. As McCain waxed poetic about Sarah Palin, he gestured with his dominant left hand, saying, "She has ignited our party and people all over America." While Bush was talking about social security ("They will continue to get their checks"), he gestured with his right hand. The hand politicians use to gesture with seems to have unexpected communicative value, providing voters with a subtle index of how the speaker feels about what he is saying. Just look at what Obama said in 2008 about health insurance while he gestured with his left hand: "You can keep your health insurance." Four years earlier Kerry made positive statements about the same subject using his right hand: "You wanna buy it, you can."[6]

Casasanto opened wide a window into understanding people's

emotions: it's not just about what people say but what they are doing with their hands. Gestures reveal a speaker's feelings about what he is saying, even when he doesn't put those thoughts into words.

Body language experts are utilized in all sorts of situations, including to prepare executives to interact more fluently with their business partners and by the FBI to help agents determine whether someone is telling the truth.[7] Often these experts focus on facial expressions and eye movements as a way to ascertain whether or not someone is expressing his true thoughts and feelings. Now we know that we can also use the particulars of hand gesture to get inside someone's head.

It turns out that right-handers agree more with messages delivered by speakers who are gesturing with their right hand; the reverse is true for left-handers. This means that knowing your audience, whether you are pitching to a right- or a left-handed client, is important. We agree more with speakers when they gesture with the hand we prefer to use, as if we were putting ourselves in the shoes of the gesturer.[8] Even though it might seem like a small detail, every advantage counts in landing that all-important deal. If you want to maximize the chance that folks will agree with your talking points, gesture about them with the hand your listener prefers to use.

It is even the case that how smooth you are when moving your arms matters. Professional poker players spend hours perfecting their expressionless "poker face," and not surprisingly, their facial expressions don't give away the cards they hold in their hands. But their arm movements do. When moving chips to place bets, poker players' actions betray the quality of their cards. Because we all tend to be smoother in our actions when we are more confident and less anxious, players judged to have smoother movements actually turned out to have better cards.[9] Our arm movements can be quite revealing. Whether you gesture when you talk or how you push an offer across the table on a folded-up piece of paper, what you do with your body reflects what's going on in your mind.

We routinely use hand gestures and other forms of body language to communicate with one another. But we don't gesture just to convey information to others; we also gesture for ourselves. We gesture while we are on the phone, even when the person we are talking to can't see us. Moving our hands helps free up brainpower. When we gesture, some of what we are working on can virtually be held at our fingertips, freeing up our mind to hold other important pieces of information.

Our understanding of the relationship between gestures and thinking was greatly enhanced by two psychologists at the University of Chicago, David McNeill and Susan Goldin-Meadow, who conducted numerous experiments demonstrating the power of gesture for a person who is moving her hands while talking. Gestures help us think and, more important, think differently than we would if we spoke without moving. One interesting aspect of gesture, Goldin-Meadow says, is that the information conveyed with our hands is often not found anywhere in the speech that accompanies it. In this way, gesture seems to reflect thoughts that speakers may not even know they have. Among students who have difficulty solving equations such as $4 + 5 + 3 = \_\_ + 3$, performance improves markedly if they are taught gestures that mimic the problem solution: grouping together the unique left-side numbers with a two-fingered "V" (that is, the 4 and the 5), and then pointing their index finger at the blank space on the right. Most striking, students are more likely to solve these types of equivalence problems if they are taught to gesture about them using the two-fingered "V" than if they are simply told to say, "I need to make one side equal to the other." Enacting the solution with their hands helps students to get a better handle (no pun intended) on the problem.[10]

It's also the case that when students are just about to grasp a problem, such as the math problem above, they often show inconsistencies between what they say about the solution and what they do with their hands. A student might, for example, group the 4 and 5 together with a two-fingered "V" and slide it over to the blank side on the right (the correct answer) while at the same time saying, "You add up 4 + 5 + 3 to get the solution" (the wrong answer). They have the correct answer hidden somewhere, but just can't seem to put it into words. When students don't move their hands, when they only talk about how to solve the problem, there aren't enough channels to convey what they know. Gesturing provides them with another avenue to express the correct answer. Having more opportunities to get the right answer—even if you don't yet know you have it—makes students more likely to learn.

How does gesture change our mind? One idea is that gestures are really just an outgrowth of how we might mentally simulate performing activities. Gestures give life to our mental scratch pads, allowing us to perform actions with our hands before we have to do them in real life or before we have thought these activities all the way through to put them into words.[11]

In one experiment, children were asked to solve the mental rotation block problem in the figure. The children tended to gesture as though they were holding the two shapes in (a), first holding the shapes apart and then moving them together and ro-

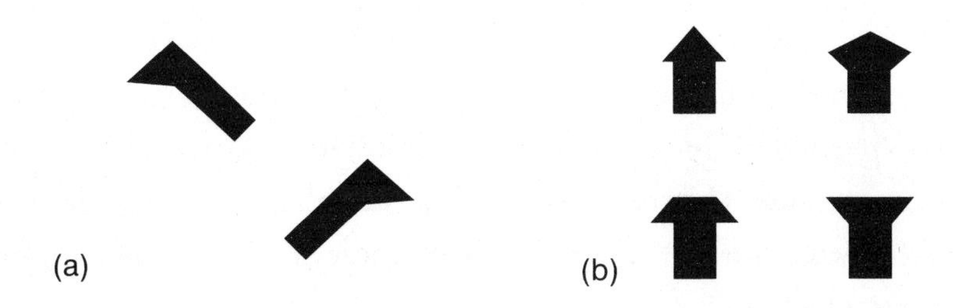

Children were asked to mentally put together the two shapes in (a) to determine which one of the shapes it matched in (b).

tating them. The gestures look just like the movements the shapes would make if they could be picked up and actually moved. The more a child moves her hands, the better she does at these space-based problems.[12]

When people explain how they perform activities ranging from hitting a golf ball to successfully flipping a crepe, they gesture. They produce motions with their hands as if they were doing the action they are talking about. Encouraging people to perform the rotation movements with the shapes, mimic the trajectory of the golf ball, or even use the "V" to understand how one side of an equivalence problem is equal to the other leads to learning. This is because what comes out in gesture adds new information to our repertoire of thoughts—information that is often more easily expressed and remembered when it is conveyed with our hands. It's also information we may not yet know we have in mind.[13]

Chemistry professors at the University of Maine have been on to the power of gesture for some time now. For the past several years, they have been encouraging students in introductory chemistry classes to gesture when they think about molecules.[14] Because a molecule is three-dimensional, students tend to be better at understanding and remembering its structure when they use their hands to represent different parts of it. It's hard to explain only in words something with a three-dimensional structure, but when we can use our hands to create a molecule in the space in front of us, we see it more clearly and remember it better.

Gestures help us learn. They also help us remember.[15] So next time you are trying to memorize information for a test or a big presentation at work, try gesturing while you practice. Gesture allows you to capture subtleties of an issue that may be difficult to convey in speech; it also gives you another way to recall the information at a later point. Just as actors are better able to remember their lines when they link them to actions, gesturing and speaking provides us with two different versions of what we are trying to commit to memory. When it is time to remember, you have two hooks (one related to action and one to speech) with which to fish out the information.

It's not just gestures. Our hand movements affect our thinking too. In this day and age, when so much of our communication is done virtually, the text messages, emails, and phone numbers we tap out serve as a primary means of conveying information. As it happens, our hand movements and the keyboard letters they are tied to have come to shape our vernacular. They even influence how we feel about the people and companies on the other end of the messages we send and the numbers we dial. How you use your body to communicate and how easily or fluently you do so shapes how you feel about all sorts of things, giving whole new meaning to the term *body language*.

Mind Control

This is the tale of the QWERTY keyboard. The name comes from the first six letters that appear in the top left row (read left to right) of the English-language typewriter. Christopher Sholes, a newspaper editor, came up with the design in 1868 after working for several years on a new and improved typewriter that would help him get his stories down faster. The device he first developed was a hassle. Like most typewriters of the day, the keys were arranged in alphabetical order and mounted on metal arms that swung up to tap the letter onto tape that was pressed into a piece of paper curled around a cartridge. When neighboring keys were depressed at the same time or in rapid succession, the arms would collide and the typewriter would jam. Because the typed letters appeared beneath the paper carriage, you couldn't immediately see if a jam had caused a mistake until you raised the carriage to inspect what you had typed.

Fed up with these problems, Sholes decided to rearrange the alphabetical keyboard he had been using and place the keys of commonly used letter pairs, like "th" and "st," far away from each other so that their metal arms wouldn't cross or collide. This helped him avoid jams and allowed him to work more efficiently. The concept worked, and Sholes sold the idea to Remington, a leading typewriter manufacturer, to replace their alphabetical layout. Frequently used letter combinations were separated on opposite sides of the keyboard and, as a clever addition, the letters in "TypeWriter" were put on the top row of keys. This way salesmen could easily tap out the brand name when hawking their product.[16]

While the QWERTY design lessened the frequency of jams, the layout still left much to be desired and did not allow typists to maximize their speed. For one thing, with the QWERTY keyboard, about three thousand English words are typed with the left hand alone (spend a few minutes typing *secret* or *exaggerated*), and only about three hundred are typed with the right (such as *milk*, *jolly*, and *hill*).[17] But most people are much faster at typing with their right hand. Not only is the right hand the preferred hand for a majority of people, but there are actually fewer letters on the right side of the QWERTY keyboard because much of the bottom right-hand row contains punctuation. Fewer letters to choose from means that it's easier to select the correct one; that is, having fewer choices leads to a quicker decision. With the QWERTY keyboard layout, people don't use their fluent right hand as much as would be optimal for typing speed.

The Dvorak keyboard is one popular alternative that gets rid of some of these issues. It's designed so that successive letters in the English language are typed by alternating hands. The distribution of words typed with the right and left hands is more even. Steve Wozniak, cofounder of Apple computer, is a big fan of the Dvorak. Some of the fastest typing speeds on record have been set on the Dvorak. According to Wikipedia, the writer Barbara Blackburn holds the 2005 Guinness World Record for the fastest English-language typing speed. She was able to maintain a speed of 150 words per minute for fifty minutes and has been clocked as fast at 212 wpm.[18] Yet despite some of the advantages of the Dvorak layout, it hasn't caught on. The QWERTY keyboard is everywhere—on computers, smartphones, laptops—across the world. And as voice communication continues to be replaced by typing and texting, the QWERTY keyboard is more dominant than ever.

Interestingly, the ubiquitous nature of the QWERTY keyboard helps explain a peculiar phenomenon: our everyday vernacular and the particular words that we like best in our language seem to be linked to how easy it is to type them. Because we tend to like what is easy to do, we prefer words typed on the QWERTY keyboard with our right hand. It's called the QWERTY effect and has been found in the English, Dutch, and Spanish languages, all of which use a similar QWERTY design. Most interesting is that the QWERTY effect is strongest for words that came into usage after the keyboard did. It's as if these neologisms are shaped by keyboard use.[19] This is one reason why LOL ("laugh out loud," typed exclusively with the right hand) and YUCKY (almost all with the right hand) are likely to stick around and why people naming new products might be wise to consider names that are typed with the right hand. A sandwich from Jimmy John's, anyone?

The QWERTY effect also seems to explain some recent trends in baby naming. By searching U.S. census data, researchers have found that the keyboard may be influencing what people are naming their babies. Names with more letters from the right side of the keyboard have increased dramatically in popularity since the dawn of the home computer and the widespread use of the QWERTY keyboard, and new names coined since the 1990s (think Lileigh) tend to have more letters from the right side than those coined before keyboard use was so ubiquitous.[20]

Our body affects our mind in unusual ways. Even when we are not performing a particular action, it seems that our mind is looking to our body for clues about how we should feel and even what we should like. Given that we spend so much time at our computers, the QWERTY keyboard is having a big impact on how we use language and even what language we prefer. In short, we like things that our body does easily.

Take the following two columns of letter pairs:

Column 1	Column 2
FV	CJ
VF	GK
BF	TK
FG	CM
FB	EJ
VR	VK
GF	BK
TF	FK
JY	JC
MJ	KB
JH	KR
HJ	KV
YJ	JD
MH	MC
UJ	KT
UH	HC

Which column of letters do you like better? There is nothing special about the letter pairs; they are fairly uncommon in the English language, and they don't rhyme. They also don't form any common acronyms or initials. So, on the surface, there doesn't seem to be a good reason to prefer one letter column over the other. But my research group and I have found that people do have a preference, at least if they are skilled touch typists on the QWERTY keyboard. Skilled typists overwhelmingly prefer the letter pairs in Column 2. Why? Just as words typed with the right hand are easier to type than those typed with the left hand, the letter pairs in Column 2 are easier to type than those in Column 1 because each letter uses a different finger and hand. As anyone who types quickly knows, typing two letters in close succession with the same finger, like the letter pairs in Column 1, is difficult because you can't start typing the second letter until you have finished typing the first. It's more fluid to type a series of letters using different fingers and hands because you can hit the keys almost simultaneously. My colleagues and I have found that just seeing letters on a computer screen revs up the motor cortex of skilled typists. We mentally simulate typing the letters and this simulation provides us with information about how easy or hard typing them is going to be. Since we tend to like what is easier to do, we prefer the letter pairs in Column 2—even though we don't usually know why.[21]

Our body controls our mind, but sometimes our body doesn't let our consciousness in on the fact that it's doing so. This subtle form of mind control isn't limited to the QWERTY keyboard. When people dial a number on their phone, for instance, the letters corresponding to the numbers they are dialing automatically seep into mind. Dialing 5683 subtly brings up the idea of LOVE, and dialing 75463 makes SLIME more likely to pop into mind. Businesses that create meaningful phone numbers are on to something. When people perform actions, the outcomes of their actions are activated mentally—even if they are not aware of it. Because the same keypad on a phone is often used for both dialing numbers and typing messages, dialing a number simultaneously brings to mind both numbers and letters.

People actually prefer dialing numbers that imply positive words (37326, DREAM) over numbers that imply negative words (75463, SLIME). People also prefer companies whose phone number corresponds to words related to their business. People prefer a dating agency with a phone number that corresponds to the word *love* and a mortician whose number spells the word *corpse* to companies whose numbers are not related to the product they are peddling.[22] This is true even though we are not necessarily conscious of the numbers' association with particular words.

The body drives the mind in subtle ways. If simply dialing a phone number can put thoughts in our head, how else does our body exert subtle mind control?

Grocery Store Choices

David Rosenbaum's discovery came to him during a meal at a restaurant. He was watching the waiters tend to the tables around him when he was struck by how they handled the water glasses of the patrons they were serving. Most of us wouldn't give the actions of our waiters a second thought, but Rosenbaum isn't most people; he runs the Laboratory for Cognition and Action at Pennsylvania State University, where he studies how people plan and control their body movements.

The water glasses on the tables were all set upside down. Rosenbaum noticed that when the waiters picked up a glass to fill it, they didn't do so in a random way. Rather they flipped their wrists and grabbed the glass with their thumbs facing downward. This initially awkward hand position had a real advantage. It allowed the waiters to easily hold the glass when they turned it over, poured the water in, and set it down. In other words, how the waiters grabbed the glass wasn't a function of its shape but was based on what they intended to do with the glass.

Intrigued by the waiters' behavior, Rosenbaum went back to his lab to run an experiment. He wanted to know whether or not we all grab objects in a way to most easily use them. Sure enough, this is exactly what he found. People pick up lightbulbs differently than they pick up tennis balls, because they usually do very different things with them. We pick up a bottle differently when we are planning to drink out of it than when we intend to throw it across the room. The way a waiter grasps a water glass depends on whether he is planning to fill it with water or put it in a tray to be cleared. Grasping the upside-down glass with your thumb turned downward means that it will take only one easy flip of the wrist to be able to fill it with water—what Rosenbaum calls the "end-state comfort effect."[23]

It turns out that monkeys also grab objects with their function in mind. When cotton-top tamarins—small monkeys that live in South America and are used in benign lab experiments—want to get a tasty marshmallow that is wedged into a champagne glass, they grab that glass differently depending on whether the glass is positioned upside down or upright. When the glass is upright, the monkeys grab it by its stem, with their thumb pointing up. But when the glass is inverted, the monkeys grab the stem with their thumb facing down, just like the waiters Rosenbaum observed.[24] That way, the monkeys will need only one flip of the wrist to get the sweet treat out.

How people and monkeys grab objects also influences how much they will come to like the object they grab. In a study my Human Performance Lab recently conducted we seated undergraduate volunteers at a small table where we had placed two different kitchen utensils: a wooden mixing spoon and a rubber spatula. We were particularly careful about how we placed the objects. Sometimes both objects were positioned so that the handle was closest to the volunteer and they were easy to pick up by the handle. Sometimes the part used to mix or flip was facing the volunteer, who had to reach around and contort her wrist in an uncomfortable way in order to grab it by the handle. Keep in mind that we didn't ask the volunteers to actually use the objects for cooking or mixing. Their only task was quite simple: to pick up the utensil they liked better.

Similar to what Rosenbaum observed with the waiters, the volunteers tended to pick up the objects by the handle, as if they were going to use it. But more interesting is that people tended to prefer the object that was easier to grasp, the object whose handle was facing them.[25] This is an example of physical fluency—the fact that objects that are easier to handle tend to be preferred. We automatically bring to mind how we would hold an object, and this ease of interaction tells us whether we will like it. That's why the right-handed majority tends to find objects on the right side of a space to be more pleasant than those on the left (presumably because they more fluently interact with objects on their right). When something is easier to manipulate, we like it more.

This means that subtle changes in the placement or packaging of products can have big effects on people's desire to buy them. It is well known that the way products are organized in store aisles and whether they are at the entrance or the exit to the store can have a striking impact on what people buy. But less attention is focused on how our fluent interactions with a product might change, say, according to how easy it is to carry. Certainly companies like Proctor & Gamble, which makes everything from hair care products to Tide, and beverage companies like Coca-Cola have tuned in to the idea that our body's actions can influence our mind. How easy it is to grasp a product influences consumer choice. Just think about the evolution of product packaging. Most liquid detergents now come in bottles with handles. This is also true for large bottles of milk and Tropicana orange juice. Several years ago the two-liter Coke bottle got a curvier look that facilitates picking up and pouring; coincidentally, sales rose over those of its closest rival, Pepsi.[26] This is no accident. Packaging that affords easier carrying might subtly push people to purchase larger quantities of a product.

In the wake of Coke's sleek new bottle, PepsiCo decided to hire a chief design officer, Mauro Porcini, who was 3M's first chief design officer. PepsiCo plans to invest $600 million in advertising to grow the market share of its leading brands, such as Pepsi, Gatorade, and Doritos.[27] Certainly some of that money is going to go into thinking about the ease and fluidity with which people handle the products. When something is easier to pick up, all else being equal, we like it more.

The first time this really dawned on me I was standing in the aisle of our nearby Target lugging my one-month-old daughter in her portable car seat. It had taken me almost a month to get up the courage to venture to Target on my own with my newborn. What if she started screaming, spit up, or both? There I was, standing with the baby in one hand, trying to decide which diapers to purchase. I automatically reached for a small package of twenty-four diapers with my free hand, easily balancing my newborn and the nappies. Somewhere in the back of my mind, however, I realized that I probably needed a bigger package of diapers, unless I wanted to test my luck with another solo store run in a few days. But I had a bad feeling about the larger package. It wasn't the price (the bigger pack was clearly more economical than the smaller pack) nor the look (the packaging was identical), but something pushed me away from the super size. Only after I started doing research on how our body influences our preferences did I realize that it was probably the simple fact that I would have a hard time carrying my newborn and the bigger diaper pack that swayed my choice to buy less. Our motor system unconsciously cues us in to the possible outcomes of our actions—even before we have completed them. Our body has a surprising influence over what we buy.

Even the type of basket we use in the grocery store can affect our purchasing habits. Picture yourself in a grocery store. You have two choices: to grab a basket or to push a cart. How many times have you opted for the basket, sure you are going to pick up only a few things, but by the end of what was supposed to be a quick run into the grocery store, you find yourself dragging a heavy basket through the checkout line? One reason you may end up with so many items you hadn't initially intended to purchase is that you tend to flex your arm when you carry a basket.

Flexing your arm and moving it toward you is something that routinely happens when you are trying to obtain an object. A lifetime of associations between flexion and gratification means that, when you flex your arm, you are more likely to want to satisfy your urges, to give in to your desires. Bringing your arm close to your body sends a subtle signal to your brain that it is okay to go for what is pleasurable. Extending your arm away from you signals the opposite. When you are in a flexed, "approach" frame of mind, you like easy gratification: you act and think in the short term rather than the long term. This frame of mind influences your shopping behavior. Whether you carry a basket on your arm (and flex your arm) or push a cart (and extend your arm) influences which products you buy.

In a recent series of studies in the Netherlands,[28] a group of researchers examined whether basket users were more likely to purchase vice products, like a candy bar or other junk food, than those who pushed carts. They inconspicuously tracked customers, selected at random, in a hypermarket (the Dutch version of a supermarket) from the time they entered the store until they left. The researchers recorded the customers' paths through the store, what they bought, and whether they were using a basket or a shopping cart.

Of course, people enter stores for different reasons and go to different parts of the store to buy the products that most interest them. And being in one part of the store could encourage more frivolous spending than being in another; think about the temptations in the snack aisle. So in order to further constrain their observations, the researchers compared the shopping behavior of people who had a basket and those who had a cart only for purchases around the cash register, where chocolate bars, candy, and gum lurk—those products that can be immediately consumed for instant gratification.

Not surprisingly, the researchers found that people who used baskets spent less time in the store, spent less money, and also purchased fewer items than people who used carts (about eleven items for basket users, compared to thirty-two for cart shoppers). However, while just under 5 percent of people with shopping carts purchased vice products, 40 percent of the basket shoppers did. Customers flexing their arms while holding a shopping basket seemed to gravitate toward products that offered immediate pleasure.

Admittedly, there could be many reasons why basket holders showed different shopping patterns than cart pushers. Perhaps people who went in with a cart were focused on the long term, stocking up for the future, and were wary of vice products that offered satisfaction in the short term. To control for these and other possible differences between shoppers, the researchers conducted a second, more controlled study. They invited volunteers to shop in a supermarket they had created. The volunteers received a shopping list with a dozen different types of products to buy—for example, meat, veggies, snacks—and were asked to choose one product from each category during their laboratory shopping trip. To do this, they were given either a basket to hold or a shopping cart to push.

In this simple study, the researchers again found that people who held a basket were more likely to pick vice products over more virtuous ones; for instance, people with the basket chose Twix and Mars Bars over apples and oranges for their snack pick. Indeed the odds of choosing the vice over the virtue were three times larger in the basket condition. How we hold our body can change what we buy, and preferences for immediate gratification increase when our arms are flexed than when they are extended.

Our body is far from a passive machine, carrying the outputs and orders our brain sends about how to act. As the researchers involved in the supermarket study suggest, "People's bodies hack their brains." How we move and even contort our body has an impact on our thoughts, the decisions we make, and even our preferences for particular products. Here are just a few other examples that come from the supermarket study:

Slot machines are designed to elicit instant gratification. Pulling levers may lead to more gambling than pushing them away or even pressing a button. The lever is also on the right side of the slot machine, which, for the mostly right-handed population, is associated with good things and also might increase the amount of money a person is willing to gamble.

Pulling open, rather than pushing, a door to enter a store might lead to purchases of vice products that provide immediate satisfaction, something that owners of ice-cream shops and liquor stores may want to keep in mind. Pulling open the door, much like flexing our arm while carrying the shopping basket, involves bringing our arm toward us, which can put us in an "easy gratification" state of mind.

Our body and the actions we perform influence our thinking and reasoning in powerful and highly predictable ways. We can learn how to notice these influences in order to fully understand how our mind works and appreciate its relationship with our body.

CHAPTER 6

Shoes, Sex, and Sports

USING OUR BODY TO UNDERSTAND OTHERS

Mind Reading

Since the time of Aristotle, philosophers have argued about whether the location of our mental abilities is the head or various other parts of the body, such as the heart. Franz Joseph Gall, a nineteenth-century German philosopher of medicine, developed cranioscopy, or phrenology, believing he could study what was going on inside a person's mind by looking at the shape of her skull. Gall argued that different parts of the brain carried out distinct mental processes—for instance, feelings of self-esteem, hope, and language—and that the subtle differences in the shape of the skull under which different parts of the brain lay could be used to infer attributes such as intelligence and moral character. As his work became accepted, it wasn't uncommon for job applicants to be sent to the local phrenologist to have their skull read as part of the hiring process, so that employers could expect their new employees to have intact concentration and conscientiousness modules.

Gall published his tome on phrenology in 1819. The English translation of the title is *The Anatomy and Physiology of the Nervous System in General, and of the Brain in Particular, with Observations upon the Possibility of Ascertaining the Several Intellectual and Moral Dispositions of Man and Animal, by the Configuration of Their Heads*. This title would never fly today—it's way too long to Tweet!

In the early 1800s a lot of people believed phrenology to be a true window into the mind.[1] The Church, however, rejected the idea that there were physical manifestations of traits such as hope or self-esteem. Others called it "bumpology," mocking the idea that one could intuit the contents of a person's mind from the shape of her skull. Scientists called the work pseudoscience and accused Gall and his followers of looking exclusively for evidence that confirmed their beliefs and ignoring findings that didn't.

The phrenology movement was rife with what psychologists today call "confirmation bias."[2] Even though it's no excuse, it's actually quite easy to fall into a confirmation bias trap. Allow me a short digression to prove the point. Take a look at the following word problem:

> You are given four cards with "A," "D," "4," "7" on one side and a rule: "If a card has a vowel on one side, then it has an even number on the other side."
>
> Your job is to test this rule to determine if it is valid. The question is: Which cards do you need to turn over to determine if the premise holds?

If you picked cards "A" and "4" you are in good company. Many people do this, but, like phrenologists, you are falling prey to confirmation bias. You do need to turn over the "A" card to see if the rule holds—there should be an even number on the other side—but it doesn't matter what is on the other side of the "4" card. There was no rule about cards with consonants on them. Perhaps they too have even numbers on the other side. What you need to do, and what most people fail to do, is to try to disconfirm the rule. To do this, you have to turn over the "7" card. If this odd-number card has a vowel on the other side, then the rule can't be true.

Most people don't go around looking for information that disconfirms their beliefs. And, this was certainly true for phrenologists. However, by the mid-1800s the lack of rigor in phrenology's claims became clear, and its popularity waned.

Not all of Gall's premises were completely off base. Modern neuroscience has certainly found evidence for specialization of brain function. Language, for example, appears to have some fairly localized centers in the brain. Using techniques such as fMRI, scientists are able to peer inside people's brain while they are speaking and have confirmed that distinct neural real estate is devoted to communicating with others and understanding what they say. Yet through this work, neuroscientists have also come to realize that the meaning of what we process is not limited to one piece of brain tissue but is distributed throughout the brain. For example, when we need to understand language about the world, we call upon the parts of the brain that support our actions and interactions—even when we aren't physically moving at all.

You may have heard the phrase "Cells that fire together, wire together." It's the simplified version of biologist Donald Hebb's striking discovery in 1949 of how malleable the structure of our brain is. Hebb found that brain cells that are repeatedly active around the same time tend to become "associated." In other words, activity in one neuron helps bring about activity in the other. Known as Hebbian learning, as cells excite one another over and over again, there is some growth or metabolic change across the connections between the cells that makes the cells more efficient at triggering each other. In the context of ascribing meaning to language, when a word is frequently encountered in the context of a particular action, hearing the word triggers activity in motor areas of the brain, which helps give rise to understanding. We understand many utterances because motor areas of the brain that would be used to do what we are hearing about get involved in making meaning out of the sounds.

This is certainly true for simple verbs such as *lick, kick,* and *pick.* Understanding their meaning isn't driven by a language minicomputer located deep inside the brain that makes sense of these words all on its own. The actual areas of the brain used to enact the body movements are also important. To understand these verbs we exploit the motor areas of the brain we use to perform these actions.[3] The word *grasp* gets its meaning because we can associate it with the grasping movements we perform; the association of the verb *give* with the act of giving grounds the meaning of this utterance in action. Even when you talk about something abstract, such as *giving your boss an idea,* the motor systems that control handing an object to someone get involved.[4]

You can think of language understanding as a mental simulation of action, using many of the same brain systems that we use to actually act in or perceive the world. This entails that words give rise to action, but also that, when we perform actions, such as turning our hand clockwise, we have an easier time understanding sentences with these actions in them: "Jessie turned up the volume."[5] When actions and words (or even phrases) are repeatedly paired together, you can't help but trigger one when the other occurs. And the more actions and words mingle together, the more fluent and deep is our understanding of language. Of course, this also implies that, when you have a disruption in the motor system, language—and especially language about action—will be impaired.

A patient was admitted to a hospital in England, in late January 2000. A few months earlier, his wife had noticed that he had become consumed with the idea that something bad was going to happen to him. At first, she didn't think much of her husband's paranoia because he had always been a little anxious and prone to premonitions of doom. But his delusions had worsened, and in the weeks leading up to his admittance to the hospital, he constantly worried that harm would befall him at any minute.

The hospital has one of the best neurology units in England, and within a few hours of the patient's arrival, a team of doctors started a general workup on him and found that his movements were slow. He was on olanzapine, an antipsychotic medication intended to control delusions; one side effect of the medication is motor difficulty, so his motor problems alone weren't unexpected. However, a scan of the structure of his brain revealed atrophy of the frontal lobes. He performed poorly on many of the mental tests he took and was also having trouble with his speech. When given sixty seconds to name as many words as possible from a given category, such as cars or fruits or words that begin with the letter "T," he could come up with only two or three items for each group.

Over the next six months, the patient continued to show signs of decline. At each return visit to the hospital, he was slower than he had been the time before and his speech more slurred. At some point all he could say was "yes" and "no"; then, one day, he lost his speech completely. He was still able to communicate in a limited manner using facial expressions and gestures, but he could no longer talk.

A group of neurologists at the hospital had been following patients for the past few years who had presented with very similar symptoms, and they became interested in this patient's case. In particular, his rapid motor and language declines were characteristic of motor neuron diseases, a group of progressive neurological disorders that destroy the cells that control voluntary muscle activity involved in speaking, walking, breathing, and swallowing.[6]

Usually when we want to execute a movement, a message from motor neurons in the brain is sent to the spinal cord. From there, the particular muscles necessary for carrying out an action get a directive to contract and act. When disruptions occur in these signals, the muscles don't work properly; they slow down, accompanied by stiffening and twitching. Eventually the ability to control movement is lost altogether. This loss of movement is devastating, but the biggest problem for people with motor neuron disease is that they have difficulty swallowing. When you can't swallow normally, it's hard to prevent inhaling foreign substances into your airways. Patients with motor neuron disease often die from aspiration pneumonia, an inflammation in the lungs that happens when food, vomit, liquid, or spit are inhaled.

Up to six out of every 100,000 people are affected with motor neuron diseases. The physicist Stephen Hawking has a form called amyotrophic lateral sclerosis (ALS), or Lou Gehrig's disease. Former New York senator Jacob Javits also had a motor neuron disease.[7]

Neurologists at the U.K. hospital devised a series of tests to better understand the problems of patients they suspected of having motor neuron diseases. In one test, patients were asked to match words such as *shoe* or *eating* with pictures that depicted the meaning of the words. Interestingly, patients performed disproportionately poorly on matching verbs to depictions of action.

In another assessment, the Pyramid and Palm Trees Test, the neurologists gave patients a picture of a pyramid followed by a picture of a fir tree and a picture of a palm tree. The goal was to choose the tree that belonged with the pyramid (the palm tree). Finally, the Kissing and Dancing Test depicted actions rather than objects. Patients might see a picture of a hand writing a letter, followed by pictures of a hand typing and a hand holding a spoon and stirring coffee. Because writing is more closely related to typing than to stirring, the appropriate match here is the typing hand. There were no differences in performance across the two tests for normally aging adults. However, the patients diagnosed with motor neuron disease did worse on the Kissing and Dancing Test than the Pyramid and Palm Trees Test.

In most languages, verbs tend to be more difficult to understand than nouns, likely due to verbs' greater grammatical complexity. This is especially true in languages such as English and Italian and less so in languages with complex noun constructions such as Greek and the Slavic languages. However, patients with other forms of damage or degeneration to their brain (for example, Alzheimer's disease) don't show this special trouble with verbs; only people with motor neuron diseases do.[8] Why? A possible reason is that the dysfunction of the motor system impairs not only actions but their linguistic correlate—verbs—as well. When you don't have the systems to act properly, understanding action language is hampered.

The patient admitted in January 2000 to the U.K. hospital died less than two years after the initial appearance of his symptoms. A postmortem examination confirmed the diagnosis of motor neuron disease. Similar to other patients who also died from the disease, the patient showed atrophy of the brain stem and spinal cord as well as a wasting away of the premotor and motor cortex.

Around the same time that the neurologists at the English hospital were discovering a link between language and action via their patients diagnosed with mirror neuron disease, the neuroscientist Friedemann Pulvermüller was also making important progress in understanding body language. For several years, Pulvermüller had been interested in what happens in the brain to precipitate impairments in people's ability to speak and understand. He was particularly fascinated by the fact that the language problems that often arose in the wake of a stroke seemed to accompany impaired motor abilities.

To understand Pulvermüller's work, it's important to take a closer look at the makeup of the motor system. The bit of neural tissue that is the motor cortex sits on the outside of the brain and straddles both hemispheres. Its role, at a most basic level, is to translate plans to act into actual actions. The nerve cells that innervate the motor cortex are organized in such a way that specific areas control particular body parts and more importance is given to the body parts that do the most work. For instance, the fingers, and especially the thumbs, have a disproportionately large representation in the motor cortex. Most people can flex and extend the ends of their thumbs fairly easily, but it's a bit harder to make analogous movements with any of their other fingers. The difference is due, in part, to disparities in the amount of brain tissue devoted to the thumb and the rest of the fingers. The thumb has more neural real estate.

It's possible to make a map of the body by documenting the connections of particular body parts to the motor cortex. Called a somatotopic map, the resulting image looks like a disfigured human with a disproportionately big face, lips, and hands compared to the rest of the body. Because of the fine motor skills these particular parts of the body need to perform, they take up a larger part of the brain's map.[9]

Somatotopic maps were first developed in the 1950s as a byproduct of a technique used to treat epilepsy.[10] At the time, a common treatment for uncontrollable epileptic seizures was to open up the patient's skull, locate the tissue from which the seizures emanated, and destroy the nerve cells there. Before the operation, however, neurosurgeons used electrical probes to stimulate different parts of the brain while the patient was conscious on the operating table so that they could observe which parts of the brain were most in control of which functions. That way

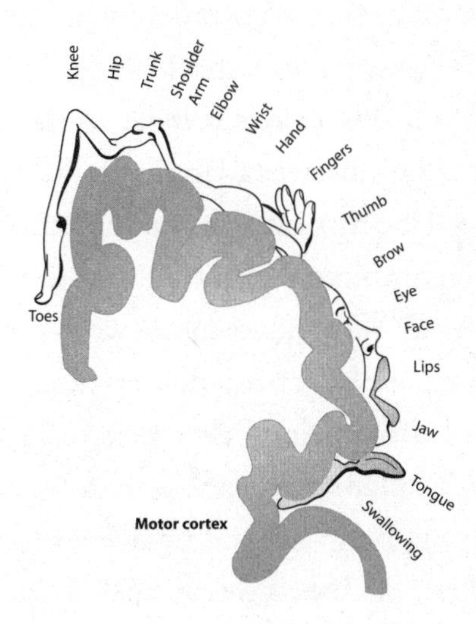

Illustration of the somatotopic organization of the motor cortex.[11]

they could target areas for removal that would cause the least disruption after surgery. This stimulation technique allowed for the creation of somatotopic maps of the motor cortex, showing the brain's connections to the rest of the body.

These body maps help to make sense of some interesting phenomena. Ever wonder why it's so nice to get a foot massage? It may be because the areas of the brain that represent the foot and those that represent the genitals are located close together. To the extent that being adjacent to each other tends to facilitate cross-talk among neurons, exciting one area may spill over and stimulate the other. This proximity of the brain's areas for the genitals and the feet may also explain why some people have foot fetishes and even shoe obsessions.[12] Although we will never know for sure unless Imelda Marcos donates her brain to science, it seems likely that these two regions are closely connected in the brain of the former first lady of the Philippines. Some neuroscientists have argued that the foot and genital areas on the brain's body map are more closely related in women than in men,[13] possibly explaining the female sex's fascination with footwear.

Pulvermüller and his research team used somatotopic body maps to study the link between language and action. While volunteers' brains were being scanned, the volunteers were instructed to perform simple movements of the hand, mouth, and feet. The volunteers also read verbs associated with moving these body parts. Moving their limbs activated areas along the motor cortex that connected to the moving body parts, and hearing verbs related to these movements triggered activity in many of the same (or closely adjacent) brain areas. Leg areas of the motor system turned on when people heard words such as *kick*; arm and hand areas were activated by words like *pick*. Action words related to the face, like *lick*, activated brain areas involved in controlling tongue movements. Most striking, Pulvermüller found that motor involvement in language processing happens extremely quickly, within just a few hundred milliseconds after a word is heard. The speed of this motor activation suggests that our action systems get involved right away when we have to understand a word, in the initial meaning-making stage.[14]

Pulvermüller's discovery has important implications. For starters, it locates one neural source of understanding: the brain areas that are used to move are also used to understand language—at least verbs. As the philosopher Ludwig Wittgenstein put it, language "is woven into" actions.[15] Most important, if the motor system helps give rise to language understanding, then language problems arising from, say, a stroke, might be alleviated by stimulating brain areas important for action. Repairing the motor system could actually help facilitate the recovery of language.

An estimated 15 million people suffer from strokes each year. Sometimes called a "brain attack," a stroke happens when blood flow to the brain is interrupted.[16] Roughly one-third of stroke victims develop problems with language, called aphasia. For some people, these language problems improve over time, but many experience chronic, long-term communication impairments. Unfortunately, treatment for chronic aphasia is limited, and people usually give up therapy long before their communication problems have been solved. Pulvermüller's work is changing this; based on his language research, he has helped pioneer a novel treatment for aphasia that is rooted in action. In his therapy, language is practiced in the context of actions. Stroke patients exercise basic as well as more advanced language skills, such as making a request or answering a question using flash card prompts as cues. The language is practiced along with the relevant actions.[17] And it's the action that seems to be doing some of the heavy lifting in helping people relearn language, even those who have had chronic aphasia for several years after a stroke. Pulvermüller's action therapy helps the brain link language with action—processing that is often disrupted after a stroke but that we now know is imperative for understanding.

Action therapy can also help healthy folks, especially when it comes to learning and understanding foreign languages. When listening to someone speak in an unfamiliar language, we often can't tell where one word ends and another begins; sentences flow together, sounding like one big word. Linguists have hypothesized that one reason we have a hard time making sense of foreign languages is because we have never made the mouth movements needed to produce the sounds. When you practice producing a properly accented foreign language, rather than just listening to someone speak, your motor practice facilitates your understanding. Having experience producing foreign words yourself helps you understand the language. Even after a relatively short amount of practice, a hundred sentences or so, you can see the benefit.[18] Our understanding of language is grounded in action, particularly those actions we can perform fluently ourselves.

Taking It Literally

One of the most amazing things about language is that we can use it to communicate literal things and actions as well as abstractions that we can't see or do. Since our ability to understand language is based on connecting to actions the events about which we read and hear, how do we make sense of things we can't see, hear, or touch? For example, how do we comprehend emotional concepts such as "giving" or metaphors such as "up in arms." The answer is that our body takes these ideas literally.

Think of a boy trying to break up with a girl. They may have been fighting a lot; their relationship has been bumpy, and the boy wants out. The girl thinks they can make it work, but to the boy the relationship is over. At a coffee shop one afternoon, in a public place where he hopes the backlash will be minimized, he tries to *cut things off*. "We're at a crossroads in our relationship," he says, and explains that he thinks they are "*headed* in different directions." Abstract concepts such as love are often described metaphorically, in terms of something tangible, like a road along which two lovers are traveling. The way we come to understand the abstract is by grounding it in the literal. As his girlfriend breaks into tears, some of the bits of her brain used to enact movement may actually be registering the metaphor and simulating movement away from her boyfriend.[19]

Admittedly, many metaphors are tied to action, so it might not be surprising that we would come to understand a variety of abstract ideas by grounding them in our physical world. For instance, sayings such as "grasping the concept" and "kicking the bucket" are abstract ideas but include the nonabstract verbs *grasp* and *kick*. However, our body is also involved in the processing of language with less clearly associated physical acts. When people are asked to respond to a sentence such as "Travis phoned you with the news" or "You told Liz the story" by pulling a lever toward them if the sentence makes sense, the content of the sentence affects how long it takes them to respond. People are faster to judge that sentences about receiving information make sense when they have to pull the lever toward them. The opposite is true too. If people are instead asked to push a lever away from them when a sentence makes sense, it takes them longer to do so when the sentence is about receiving news compared to when the sentence is about giving news to someone else. This is because even the abstract idea of transfer, of giving, is grounded in the actions of giving and receiving ourselves. And when there is a match between the abstract idea and our actions, performance improves. Exchanging ideas is seen as an extension of exchanging objects and therefore is linked to many of the same motor and perceptual processes. During language comprehension, the motor system is activated when language implies that something is being transferred, whether or not it is physically moving.[20]

We understand the abstract by mapping it onto the tangible. Just think about the concept of time. We often talk about time as space and use spatial metaphors to do so, for instance, "He moved the meeting forward two hours." We take the abstract concept of time and play it out concretely in terms of our bodily movements. Sentences such as "I am looking forward to our date on Friday" and "I am thinking back about dinner last week" show how we evoke something with parameters or boundaries, like space, to make sense of something that is harder to comprehend, like time. Yet we rarely talk about space as time. It would be odd to say "This place is a long time" to refer to a large amount of space. But you might say "The time is short" to refer to an impending deadline. When people standing in a lunch line were asked when a meeting scheduled for Wednesday would take place if it were "moved forward" by two days, those customers who had moved farther forward in line were more likely to respond that the meeting would be on Friday (rather than Monday) than customers farther back in line.[21] How we move through space influences how we think about time. Because we can physically navigate through space but not through time, we tend to use the former to think about the latter and not vice versa.

Even when people recall past events or project events into the future, their body seems surreptitiously to play out the metaphor of time as space. When people think about events in the past, they lean backward slightly, and when imagining the future they lean forward. These are small shifts, only several millimeters in one direction or the other, but nonetheless they exemplify our inclination to translate time into how our body moves through space.[22]

Experience Matters

> LeBron starts off the game with a crippling slam off a Thunder turnover.
>
> LeBron dribbles the ball to the top of the key, drives the lane, and finger-rolls over Kevin Durant for two.

If you are a basketball fan, and especially if you play in your free time (or even if you laced up your high-tops once upon a time in high school), these sentences make perfect sense. Your spectating brain is also likely playing off what you are hearing as if you were LeBron. That's right, to some extent your brain thinks that you are part of the Miami Heat on the court. Of course, you might also be an Oklahoma City Thunder player at one point in the broadcast—not so fun when you are playing against the Heat in the 2012 NBA Championships.

We understand words like *throw* because we have learned to pair throwing actions with throwing words. Words like *throw* or *dribble* derive their meaning, in part, because of our previous experiences. Does this mean we can't understand what we are reading if we have never done the action we are reading about? Not necessarily. If you have never used chopsticks, you can still understand the sentence "*Mila picked up the wonton with her chopsticks*" by extrapolation to familiar activities like eating with a fork or holding a pencil in your fingertips. Nonetheless being able to call upon an experienced motor system can aid your understanding.

Knowing that your experiences acting in the world influence what you understand helps explain why athletes and sports fans alike mimic plays they see on the field or on television or hear on the radio. In a study conducted in my psychology laboratory at Miami University several years ago, we took a look at this mimicry. If you follow professional football, you probably know that Miami University is not in Florida, but in Oxford, Ohio, a small rural town about forty-five minutes northwest of Cincinnati. You probably know this because Ben Roethlisberger, the quarterback for the Pittsburgh Steelers, played college ball there. Miami also houses a top-ranked ice hockey team. When I was on Miami's faculty, Brian Sipotz, one of the team's star defensemen, worked as a research assistant in my psychology laboratory, so I saw them play often.

Brian was convinced that playing hockey made him a different kind of hockey fan from his nonathlete friends; he believed he was better able to comprehend what he witnessed on the ice. When he watched a game he felt almost as if he were playing, involuntarily twitching and moving as he assumed the identity of the player with the puck. This feeling wasn't limited to watching a game: Brian felt just as much a part of the action when he audio-streamed his favorite NHL match-ups on his computer. Could Brian's experience on the ice have changed how he understood the athletes he followed as a fan? We decided to test Brian's hunch by inviting his hockey teammates and another group of guys who weren't athletes to have their brain scanned while they listened to a simulated audio broadcast of a hockey game.

Afterward we quizzed everyone to see how well they had followed what they had heard.

The hockey players' motor system (specifically the premotor cortex) came alive when they listened to the hockey broadcast. This wasn't the case for the guys who didn't play hockey; when listening to the broadcast, the fold of cortex responsible for choreographing movement was relatively idle—it wasn't as active as it was for the hockey players. Because the hockey players were hockey players, they were able to closely simulate in their brain the behavior of the players in the game they were listening to. And, the harder the premotor cortex worked, the better people were at following the action.[23]

Our work with hockey players provides a new understanding of what happens in the minds of sports fans as they sit on the couch or in the stands, or even just while listening to a game. Their brains are playing along. They might even mimic the movements of the athletes they are watching or listening too. This mimicry might just seem like eager fan behavior, but it's actually related to the fans' own skill. When we observe or even hear about others' actions—especially if we have done something similar in the past—we aren't just watching; at least, our motor cortex doesn't just sit idly by. Rather, in our brain we act out what we are watching, as if we were one of the players ourselves.

Capoeira, a Brazilian art form that is a cross between dance and martial arts,[24] was created by the descendants of African slaves who had been transported to Brazil to farm the land and harvest sugar cane. The slaves were overworked and underfed and lacked even basic material comforts. Capoeira developed as more than a dance—it was a way to express anger and frustration and was also a fighting style that could be used if a slave needed to defend himself.

Today Capoeira is practiced all over the world and appears in popular movies and video games. The 1993 martial arts film *Only the Strong* showcases it, and the actor Mark Dacascos uses it as a way to mobilize youth in his gang-infested hometown of Miami, Florida. One of the main characters of the video game Street Fighter fights in the capoeira tradition. It has even shown up in neuroscience research as a way to demonstrate how important our own experiences are for understanding the actions of others.

When experts in capoeira watch it being performed, their brain circuits for capoeira movements become highly active. The brains of classical ballet dancers who watch the Brazilian art form don't tune in to it in the same way. Interestingly, this perking up of the motor system is really about motor experience, not simply being familiar with the dance style. When ballet dancers watch moves from their own repertoire and moves that the opposite sex performs but they don't, their motor circuits react to what they can do, not what they have watched. Having an internalized copy of what you are watching helps you understand what you are seeing.[25]

Getting your motor system into the game (or dance) has some real advantages. For example, it allows you to predict the outcomes of other people's actions before they have actually been performed. This is beneficial on the sports field as well as when you are simply trying to follow a game. Referees and sports writers could benefit from having played the game themselves.

A few years ago a group of neuroscientists from Rome conducted an experiment in which they asked basketball players, sports journalists, and folks with no experience playing the sport to watch film clips of basketball players attempting free throws. The clips were stopped at various time points throughout the shot, and people were asked to predict whether or not the ball would eventually end up in the basket. Not surprisingly, the players were better predictors. But what was most interesting is that the players showed a real advantage early on in the shot action. Even before the ball had left the hands of the shooter, the players were out-predicting the journalists and basketball novices. Experience gave the players an edge in understanding how a shot they were watching was going to end up.[26]

As people watched the shots, the scientists watched for signals from electrodes they had placed on their subjects' hands and forearms to see if there were any evidence that the body was getting ready for action. Although the motor signals being sent from all the viewers' brains to their arms and hand muscles perked up somewhat when watching the shots, it was only in the players that the extent of the motor excitement in the hand muscles controlling the pinkie finger predicted whether the shot was going to be made or missed. And there was more finger activation when the player in the film clip bounced the ball off the rim and missed the shot.

Elite athletes in other sports can predict performances too. Skilled badminton players watching films of players about to hit a birdie can predict the landing position of the shots even when they can't see a majority of the opponent's arm and racquet. Beginners, on the other hand, need all that information to make the same predictions.[27] In baseball, batters often start their swing before the ball has left the pitcher's hand because they can start predicting where the ball is going to go by the movement of the pitcher's body. An experienced brain can pick up on what others are doing, mirror their actions, and send the appropriate signals to the body so that it knows what to expect and where to move before an event has been fully played out. That's why skilled performers always seem to be two steps ahead: their brain has been rewired to play out actions before the actions have happened in reality.

The key to this ability may be grounded in a neural circuit called the forward model, which helps our brain predict the outcomes of our actions (and the actions of others) before they have happened. When we decide on an action and our brain signals the muscles to perform it, a copy of this command is created that estimates the end result of the movement. It gives us feedback from our senses before we have actually completed the action. When you move your hand from one place to another, the brain estimates its new position and what is going on before any feedback from the outside world arises. These predictions are one reason why, when you touch a hot stove, you might move your hand before you have actually felt your skin burn. You predict what is going to happen before you register the pain. When the actual feedback comes in, the burning sensation can be largely ignored if it matches what your brain has already predicted.

Scientists think they can see these forward models in action. For instance, the cerebellum, a wad of tissue on the back underside of the brain, is important for controlling the timing of movements (among other things). How neurons fire in the cerebellum of cats can be used to predict the trajectory of a moving target. The cat's cerebellum can predict the motion of the target before it lands, and likely the more experience cats have seeing things move, the more accurate their prediction.[28]

Think about a gymnast performing on the beam at the Olympics. Doing several back-to-back flips across this 3.9-inch wooden block means that she has to know exactly where she is going to land before she does so. It's the only way to prepare for her next move. This need for anticipation extends to most sports where you have to produce actions too quickly to get feedback from the environment. In tennis, players often have to start moving before the ball has left the opponent's racket; in skiing, racers have to think at least two gates ahead to have time to set up for their turns. Experts' brains must anticipate the future of actions before they have completed them so they can act fast and adjust when needed. Because of their immense amount of practice and experience, elite athletes are able to take what they see or what they intend to do and form a good picture of how it will turn out. Experience means you don't need to reason step by step through your actions or even the actions of others. Rather you are able to start to play out what will happen in your head before you have to actually perform it.

Though skilled players don't need to use the racquet to predict the landing of the birdie or use the ball to predict where it will go, they think they do. Athletes are not always aware of all the cues they use to anticipate actions. Perhaps this is one reason why the best athletes don't always make the best coaches. They can't introspect on what they do to teach it to someone new to the sport. Perhaps this is the case with Wayne Gretzky. After years of dominating on the ice, Gretzky has struggled to put together a winning team, whether it's at the Olympics or in the NHL. As a player, Gretzky seemed to anticipate what an opponent was going to do before the other player knew himself, but this ability is hard to teach.[29]

It's pretty obvious that our thoughts can drive our behavior. But a lot of what we know about the world comes from being able to move around in it. The body influences the mind too. Perhaps that's why people say you have to have had some piano playing experience to truly appreciate Stravinsky's *Petrushka* and why the most passionate sports fans were once athletes themselves. In short, what we do physically dictates what we understand mentally. What we do physically also dictates how captivated we are by what we see and hear around us.

CHAPTER 7

Tearjerkers

EMPATHIZING WITH OTHERS

The movie *Love Story* may be one of the greatest tearjerkers of modern cinema. Set on the Harvard University campus in the early 1970s, the movie tells the tale of students Oliver Barrett IV and Jennifer Cavalleri, who meet at the Radcliffe library, where Jenny is working her way through school. Despite the fact that they come from different backgrounds (Oliver is a Harvard Law student from a wealthy family, while Jenny is a Radcliffe art major from a modest, working-class family), they are instantly drawn to each other. After long walks in the park, romantic dinners, and study dates, they decide to marry, but Oliver's dad doesn't approve of the union and severs all ties with the couple, including much-needed financial support.

The newlyweds struggle to make ends meet. While Jenny works long hours as a private school teacher, Oliver graduates third in his class and takes a job at a top New York firm. When the couple tries to find out why Jenny is having difficulty conceiving, they discover that she is terminally ill. Beside himself with grief, Oliver once again finds himself in dire financial straits from the staggering hospital bills that have accrued during Jenny's illness. Swallowing his pride, he goes to his father to ask for help, but he is too proud to tell his dad that Jenny's illness is the reason he needs the money. Instead Oliver confirms his dad's accusation that the money would be used to help another woman he has gotten pregnant.

One of the last scenes of the movie is set at the hospital during Jenny's final hours, with Oliver by her side. His father discovers why his son really needed to borrow money and immediately heads to the hospital to make amends. Jenny has passed away by the time he arrives. When Oliver's father apologizes to his son for his behavior, Oliver responds with an epic line, based on something that Jenny said: "Love means never having to say you're sorry."

Even though audiences know that the movie is just a movie, many people come close to tears just thinking about *Love Story*. Movie producers bank on this reaction, the fact that our neural circuitry doesn't always make a clear distinction between what is real and what is not. Directors work hard to draw audiences into the story so that they experience the emotions of the characters as if they were their own. Movies like *Love Story* are called "tearjerkers" for a reason.

We cry while watching tearjerkers even when we know the actors are only portraying emotions because our mind largely sees the emotions as real. This is also true when we read a sad love story; we empathize with the characters as if their trials and tribulations were our own. We make sense of what we are reading or watching by bringing to the surface the thoughts, feelings, emotions—even the sights, sounds, and smells—that we encountered when we were in a similar situation. That's one reason why *Love Story* is so popular. We all can relate.

Many scientists believe that empathy, the ability to appreciate the emotions and feelings of others, is largely explained by the phrase "Monkey see, monkey do." Perhaps not surprisingly, the discovery of mirror neurons—those brain cells that respond not only when a monkey performs an action but when the monkey sees someone else perform the same action—are important for thinking about this resonance. Mirror neurons are most often talked about in terms of understanding *action*, but mirror neurons and the idea of mirroring more generally are also important for understanding others' *emotions and feelings*. Our ability to comprehend the feelings of those around us works, at least in part, by mirroring the actions and related emotions we observe. We recognize another person's emotions by mapping his behavior onto something similar in our own repertoire of actions whose emotional state we already know. By doing this, we can get a pretty good idea of what he is feeling, even if we can't see all of his emotions directly and he isn't telling us what he feels. The idea of mirroring helps explain how a direct link between seeing and feeling could come about.[1]

Consider a study conducted by a colleague of mine, the neuroscientist Jean Decety, at the University of Chicago. Decety began by asking volunteers to read short descriptions of events that were likely to elicit intense emotions; for instance, "Someone opens the bathroom door that you have forgotten to lock." He asked one group of volunteers to read these scenarios and think specifically about being in the situation themselves. He asked another group to think about their mother being in the same situation. It's hard not to cringe when you think about your mother sitting on the toilet.[2]

Decety used fMRI to peer inside the volunteers' heads while they imagined these scenarios. He saw that areas of the cortex important for the processing of emotional information, such as the amygdala, became excited when people thought about themselves in these embarrassing situations *and* when they thought about their mother going through the same ordeal. Some of the same bits of brain tissue we use to experience emotions in ourselves are reused when we think about other people experiencing the same things. Decety's work provides a clue as to why we want to cry (and some of us actually do) when we find out that Jenny is going to die in *Love Story*. The emotion centers of our brain register the turmoil as if it were our own. Just as experienced athletes are able to understand others' play and predict an opponent's next move by re-creating the movements in their own head, we resonate with the emotions portrayed by characters in movies like *Love Story* because we experience the anguish ourselves.

This merging of the self and others happens routinely. When people watch videos of others showing facial expressions of disgust, some of the same areas of the cortex come alive as when a person inhales noxious odors, say, the smell of rotten eggs. We recognize facial expressions of disgust and even words that depict disgust because we ourselves have felt disgust.[3]

Interestingly, these empathetic reactions start very early in life. One-day-old babies cry more when they are exposed to other infants crying than when silence surrounds them. And here is where the findings get really interesting: the infants cry more when they hear another baby cry compared to when they hear synthetic or artificially generated cries of the same intensity. The newborns are most affected by infant cries that are similar to theirs. Scientists believe this suggests that we may be endowed from birth with an innate capacity for empathetic reactions. We glom on to the emotional reactions of people who are most like us and experience their emotions as if they were our own. This linking of ourselves and others is really powerful; it gives meaning to the idea that "others are like I am" and forms the building blocks for empathy in later years.[4]

Given that we activate some of the same emotion centers of the brain when we are in distress and when we see or hear the distress of others—especially those who are most like us—how do we ever separate ourselves from others? It turns out that, at least early on, we don't do a good job of differentiating. Children don't usually make a distinction between their own thinking and that of other people. Early in life, we don't yet have an intact "theory of mind," the understanding that our thoughts and feelings may differ from another's. A simple but clever assessment, the Sally-Anne Test (one of the many different versions of what psychologists call a false-belief task), exemplifies this phenomenon quite clearly.

A three-year-old child is told a story about two girls, Sally and Anne. As a way to set the stage, the child is usually shown dolls that represent the girls. In the story, Sally has a basket beside her, and Anne has a wooden box. The child is told that Sally has a toy that she decides to place in her basket before leaving for a while. (When the toy is in the basket, no one can see it.) Sally leaves, and Anne goes over to Sally's basket, takes out the toy, and puts it in her box (again hiding it from view). Sally then comes back, and the child is asked, "Where will Sally look for her toy?" The correct answer of course is that Sally should look in her basket for the toy, where she left it. But, depending on the child's age, she may not differentiate between what she knows (her own mental state) and what others, in this case Sally, should believe. The child knows the toy is now in Anne's box, but theory of mind entails that she understand that her knowledge may be different from that of other people, that Sally doesn't know the toy was moved.

Most normally developing children pass some version of the Sally-Anne Test by around four years of age. Before this time, however, not being able to make a clear distinction between ourselves and others is a normal part of the developmental process. The merging of me and you early in life serves as a basis for shared feeling and understanding—a crucial component of empathy. For instance, by automatically matching his mother's behavior and his own emotional state, and not being able to tell the two apart, a child becomes highly attuned to what his mother is feeling. Because of this, he is able to form a close connection with her. Infants of depressed mothers tend to display negative facial expressions more frequently than do infants of nondepressed mothers. Because depressed mothers show more negative affect on a daily basis, their babies are more in tune with these responses and synchronize to them.[5] In the short term, this bodily mirroring helps the baby relate to his closest caretaker. In the long term, of course, having a depressed mother can have dire consequences. When children constantly mirror the negative affective behaviors of their parents, the children's own expression of negative bodily emotions sends signals to their brain about how the children feel. In this way, depression can be handed down from parent to child, with the body serving as a vehicle of transfer. A genetic predisposition for depression may certainly help explain the linking of depression in mother and child, but how a child holds his body (which is often learned from what the parents are doing) matters a lot too.

At some point during development we do learn to separate our own feelings from the feelings of those around us. Yet even as adults, we call upon our body to help us make sense of potentially emotional information that others display, and this bodily involvement has some striking consequences. Paula Niedenthal, a social psychologist at the University of Wisconsin, has spent the past several decades investigating the link between the body and emotion. Although people tend to think that affective reactions are rooted in the mind, Niedenthal has repeatedly demonstrated the important role the body plays in our emotional experiences. In one of her most compelling studies, Niedenthal asked students participating in her research to decide if a particular object, such as a baby, a slug, or a water bottle, was associated with an emotion. Unbeknownst to the students, Niedenthal had picked the objects to be either highly emotional—igniting strong feelings of joy, disgust, or anger—or not emotionally laden at all. In addition to rating the objects, the students also rated more abstract concepts, such as joyful and enraged, as emotional or not.

While wearing small electrodes placed below the mouth and above the eye, the students judged the emotionality of the objects and ideas, and Niedenthal took recordings from their facial muscles. The tiny movements, which are invisible to the naked eye, can be used to assess whether facial muscles produce frowns or smiles. The results were clear: while making their judgments about both the objects and the abstract ideas, the students reflected the corresponding emotion on their face. It took them only a few seconds to decide if a slug, for instance, was strongly associated with an emotion, but in that time their face also expressed signs of that emotion.

The emotional centers of the brain send signals to the body about how we should behave, but that pathway is not a one-way street. The muscles involved in the body postures we assume and the facial expressions we make also send signals back up to the brain, reinforcing our feelings. In a provocative demonstration of this point, Niedenthal ran another experiment in which she asked volunteers to watch videos of people whose face was morphing from one expression to another—happy to sad or angry to amused—and to indicate when the facial expression had changed by pressing a button on a keyboard.

Not surprisingly, as the volunteers monitored the faces for changing emotions, they were morphing their own facial expressions along with the videos. Niedenthal allowed some volunteers to freely mimic the facial expressions they were watching, but she asked another group of volunteers to hold a pencil between their lips and teeth, which prevented them from frowning or smiling along with the faces they were viewing, even though they didn't realize it. Subjects who were free to mimic detected the emotional change in the video faces much quicker than those who were prevented from mimicking. Our facial expressions send feedback to our brain about what emotion we should be feeling, which in turn affects our ability to understand emotions in others. As Niedenthal notes, these findings lend credence to the saying "When you're smiling, the whole word smiles with you."[6]

But what if your facial muscles aren't sending those signals? In the first chapter of this book I described how Botox prevented frowning and the negative emotional signals that go along with it. Freezing the face can just as easily prevent positive emotional information from being sent to the brain. Niedenthal has been examining the social and emotional consequences of the pacifier, an object that is a very important part of my life as the parent of a toddler. She had been wondering if the pacifier serves the same sort of function as putting a pencil between your teeth, freezing the facial muscles so that babies are less likely to form emotions on their face. Could a pacifier stunt emotional development by preventing a child from mimicking others' facial expressions, perhaps having some effect on that child's eventual empathetic reactions as an adult? Although she doesn't have a definitive answer yet, Niedenthal thinks parents need to be cautious about how often they let their child have a pacifier. Even though we often think of the pacifier as a life saver, soothing an infant in times of stress, if it prevents a child from fully displaying the emotions he is feeling or mirroring those observed in others, there may be some risks. Yes, a pacifier may help stop the wailing and forming of unhappy expressions that could turn into a depressive cycle: the child is unhappy, Mom is unhappy, and the child is unhappy in return. But it's also important to remember that, whether positive or negative, the ability to freely display emotions on our face is an important part of learning to successfully empathize with others and to fully experience emotions ourselves.

How we contort our body when we spontaneously imitate others gives us insight into other people's emotional states—a vital piece of the empathetic puzzle. Interestingly, this interplay between ourselves and those around us helps explain a surprising phenomenon: why married couples often look alike after many years together.

Married couples are highly motivated to empathize with one another; to do this, they mimic each other's facial expressions, which in turn facilitates similar emotional experiences. People who are on the same page get along better and are more likely to stay happily married. Over time, research confirms, this mimicry leads to permanent changes in how the face is shaped. In one study, more than a hundred volunteers were shown photographs of men and women taken in their first year of marriage and taken twenty-five years later, on the spouses' silver wedding anniversary. They were also shown photos of randomly matched pairs at the same ages. The volunteers were asked to judge the physical similarity of the couples. Sure enough, there was an increase in similarity among the married couples at the twenty-five-year mark, but not the randomly matched pairs. Most striking, the more similar people looked, the happier they reported they were in their union.[7] So next time you are struggling to connect with your spouse, try to subtly mimic his or her facial expression; it is likely to make you feel in synch and strengthen that emotional link that can weaken in times of strife.

My Emotions Are Not Yours

In one of the final scenes in the movie *Love Story,* Jenny is in the hospital on her death bed with her husband by her side. Doctors and nurses come and go to check on her but seem unaffected by the couple's ordeal, which never seems to strike viewers as odd. Doctors are supposed to have a level of professionalism that removes them from their patients' pain and suffering; this allows them to dispassionately make difficult diagnoses and decisions about treatment. Yet physicians also need to have some level of empathy for their patients. Doctor empathy is particularly important for successful communication with patients and is also associated with improved patient satisfaction. It even relates to a patient's tendency to comply with a recommended treatment. How do doctors emotionally connect with their patients without becoming overinvolved in a way that can preclude effective medical care?

Most doctors don't exhibit downgraded emotional responses outside of their practice, which suggests that it's experience rather than a natural inclination for emotional distance that endows medical professionals with the ability to cope with their patients' suffering. Doctors learn techniques for keeping their emotions in check and for focusing on what they need to accomplish in the situations they face, such as an emergency or a patient who is not responding to treatment as expected. Indeed people just starting a career in medicine show greater overt emotional reactions to other people's pain than those with a lot of medical experience. What's more, the place in the brain where differences emerge between how doctors and nonmedical professionals respond to the discomfort of others tells us a lot about how doctors cope with suffering.

There is a striking overlap in the neural circuits that drive the firsthand experience of pain and the perception of pain in other people. The insula, somatosensory cortex, and cingulate cortex are involved in our own experience of pain and the processing of others' painful experiences.[8] Our empathy largely draws upon a resonance between ourselves and others: we mentally simulate, and hence share, other people's emotional experiences. When doctors watch video clips of body parts being pricked by a needle, they show less activity in brain regions at the neural epicenter of the pain response than do nondoctors. But it's not that doctors' brains are less active overall when witnessing a painful event. Quite the opposite. An area smack dab in the middle of the frontal cortex that houses our ability to regulate our feelings and emotions works harder in doctors when they watch other people in pain. Indeed the more active the doctors' emotion-regulation brain center is, the less involved the bits of brain tissue that register pain.[9] Doctors train their prefrontal cortex to rein in their natural inclination to mirror other people's painful experiences.

The ability to temper our emotions, which tend to spontaneously ignite when we see others in distress, develops over the course of a lifetime. Seeing someone's hand getting slammed in a car door is a very different experience (at least for our brain) when we are seven years old than when we are in our thirties. There is a mental and physical shift in how we understand these emotional situations that depends on our age. Young children show a more visceral emotional response that is critical for determining the emotional meaning of what they are seeing. These kids might actually flinch and may even grab their hand at seeing a car door slam on another person's hand. In contrast, adults show a more reasoned and detached reaction, more similar to that of seasoned medical professionals. As we age, we get better at making sense of the feelings we see in others, including our ability to separate others' emotions from our own. The development of the prefrontal cortex, which continues well into our mid-twenties, certainly promotes this emotional meaning-making. Before this time, it is easier for the emotional centers of the brain to call the shots.

Just as doctors train their prefrontal cortex to temper their pain responses, we all develop this skill over the course of our lifetime to rein in our emotions in all sorts of harrowing or stressful situations. Math phobics perform better on a math test when they call upon some of the same emotion-regulation processes that doctors do. Likewise, people who have a phobia, say, a fear of spiders, can use strategies to put their fearful reactions in check in order to approach a tarantula—venturing closer and even reaching out and touching their eight-legged terror. How do those with math anxiety and arachnophobia do it? One technique is as simple as writing down your thoughts and worries about the negative event. Writing for even as little as ten minutes helps download those negative thoughts from your mind, making your negative emotions less likely to boil up and distract you from the task at hand.[10] In a sense, the writing helps your prefrontal cortex turn down the volume on the loud speaker of your negative reactions.

The point is that our emotions and fears don't have to get the best of us. We just need to wield tools that can help us temper these negative reactions when they bubble up to the surface and threaten to derail our ability to perform at our best. Medical professionals learn how to do it as a means to separate themselves from their patients' pain and suffering. We can too.

Think back to my colleague Jean Decety's experiment, where he asked people to imagine themselves being caught on the toilet and their mother in the same predicament. Although Decety found that many of the same emotion-laden areas of the brain came alive when people thought about themselves and their mother, there were also some differences. Most striking, activity in the sensory cortex indicated which scenario a person was thinking about. Sitting directly behind the motor cortex, the sensory cortex is the strip of brain that is responsible for receiving incoming messages from our senses—touch, hearing, smell. The sensory cortex is especially active when we think about ourselves rather than someone else, likely because when we think about ourselves we call more directly upon our previous physical experiences.

A closely related bit of brain tissue, the temporal-parietal junction (TPJ) is also important for helping us become consciously aware of whose feelings belong to whom, separating our own feelings and actions from someone else's. The TPJ receives inputs from our different senses and integrates various pieces of body-related information, helping us to form a holistic picture of how we are feeling. Because of its role as a body monitor, the TPJ is believed to be an important player in our development of theory of mind—our ability to know that our thoughts, actions, and intentions might be different from another's. Together with the sensory cortex, our TPJ signals when the feelings we have are the result of our own experiences or are just an empathetic reaction to what others may be experiencing.[11]

Abnormalities in the TPJ and surrounding brain tissue are sometimes associated with autism spectrum disorders.[12] This has led to speculation that faulty construction of neural areas that help us differentiate our own actions and intentions from those of others (or, at the very least, a problem in the signals being sent to and from these areas) contributes to autism. People with autism have problems with social interactions, especially understanding the emotions and intentions of others. If you don't recognize that another person's actions—say, a smile, frown, or grimace—are similar to those in your own motor repertoire, you will have difficulty making sense of other people's behavior.

Autism: Broken Mirrors?

Not all children who have been diagnosed with an autism spectrum disorder (ASD) show visible signs of a problem. But interactions with them, attempts to engage them in a conversation, for instance, often betray their challenges. A child may avoid your gaze, may not answer normally, and may rock back and forth or even put his head in his hands. Unlike a typically developing child, a child with ASD may not be able to accurately read your facial expression or body postures, and he won't use your social cues to understand what you are thinking or feeling. If you stick out your tongue in play, he is unlikely to stick out his tongue in imitation, as many other children would. Some children diagnosed with ASD find imitation difficult to do.

The body postures and facial expressions of others are extremely important sources of social information. They tell us about a person's emotional state, whether he is friend or foe, what he intends to do, and what actions we might take in response.

Being able to accurately perceive and recognize social cues displayed by others is key for social interaction, but not everyone can do this. Individuals with ASD (which is estimated to occur in one out of eighty-eight children)[13] have great difficulty understanding the social information—especially nonverbal cues—displayed by others. Some scientists believe that this deficit in social processing stems from a broken or faulty mirror neuron system; abnormalities have been found in sensory, motor, and related brain regions that help to initiate our own actions and ascribe meaning to the actions of others by assimilating them with something we have done in the past.[14]

To study human mirroring, researchers often use an imaging method called an electroencephalogram (EEG); the patient or subject wears a cumbersome cap full of electrodes that transmit signals to a screen to create a picture of the person's brain waves. For some time now, researchers have known that a specific component of brain waves, called a mu wave, is suppressed when we voluntarily make a movement, such as reaching out to grab a bottle. Although neurons emanating from the sensory and motor centers of the brain fire in sync when people are resting, initiating a movement actually disrupts this synchronicity; as a result, the amplitude of the mu wave plummets (termed mu suppression). Most striking, these mu waves are also blocked when we watch someone else perform an action. Just as the mirror neurons of rhesus monkeys fire when they reach out to grab something and when they watch someone else grab the same thing, your own brain waves that signal action change in predictable (and similar) ways when you act and when you watch someone else act. Given this similarity in mu wave suppression when people perceive and perform actions, researchers believe that the mu wave is a possible indication of mirror neuron activity.

In one experiment, children who wore an EEG cap were asked to grab an object and also to watch videos of other kids grabbing the same object. When normally developing children made the grab, their brain showed the same activity as when they watched others do so. However, the EEGs of children with autism signaled action only when they grabbed the object themselves. It seems that ASD kids don't always register the actions of other people, at least as an action that they might do themselves.[15]

Recent evidence indicates that we can learn to suppress our mu waves through biofeedback training. Professor Jaime Pineda of the University of California at San Diego has been exploring whether children diagnosed with ASD can learn to regulate their brain rhythms so they have better control over how they understand and react to others. Pineda has devoted his career to understanding how our brain takes in and processes information from the outside world. If you met Pineda outside his corner office in the modern cognitive science building at the university, you might not know he was a notable neuroscientist. He is very unassuming, with a soft voice, dancing eyes, and a warm smile that give off an air of creativity most folks would associate more with an artist than a scientist. But his creativity is clearly reflected in his outside-the-box research program.

In one study, Pineda recruited local San Diego children who had been diagnosed with ASD to take part in a neurofeedback training program.[16] All of the children were high functioning, with generally normal IQ and verbal skills appropriate for their age. Their parents were all members of Valerie's List, a San Diego Internet autism support group. A few times a week, over a ten-week period, the kids visited Pineda's laboratory to take part in the training, during which they wore an EEG cap that monitored their brain's electrical activity. Pineda and his team taught the kids how to control their brain waves using several different video games involving race cars, robots, and space exploration. The children learned how to use their thinking to move objects on the screen, for instance, moving a race car around a track. They went through about fifteen hours of training in total. Following the ten weeks, their parents reported positive changes in attention, interactions, and other social behaviors that often go hand in hand with autism (compared with children who didn't receive the training). If kids with ASD can learn to alter their mu waves, this could prompt development of new therapies for autism. Specifically, these games could be used to reinforce mu suppression both when a child is acting and when he is watching someone else act, positively affecting his ability to navigate his world and successfully interpret the behaviors of those around him.

This loop of behavioral change, from neurofeedback training to a lessening of the symptoms associated with autism, illuminates the power of the body and the corresponding brain signals that control action. When children change their mental patterns, their own behaviors change. Perhaps it helps kids with autism to make meaning out of social interactions when they see that their own actions and the actions of those around them are closely related. Instead of simply seeing a series of body movements when an adult playfully sticks out his tongue, a child can connect the action with meaning and recognize that the adult is thinking about playing and trying to get a playful response.

Many scientists, however, argue that there is still not enough evidence to embrace the idea that an impaired mirror system underlies autism.[17] Because autism often goes hand in hand with all sorts of cognitive and motor deficits, it is difficult to be sure that a mirroring issue is really driving the disorder. Something broader may be at work, for instance, a failure to pay special attention to people and their actions that leads to deficits in understanding social information. Though some children with ASD have difficulty imitating others' actions, some do not. It might be that children with ASD just don't know when to imitate. It's as if children with autism are not able to use social cues in order to understand how to behave.

Simply put, kids with ASD process social information differently, not giving it the preference (at least in the brain) that typically developing children do. It seems as if, in the autistic brain, social information is no different from any other sort of information that we encounter. A smile is not recognized as a signal of friendship; it is just the facial muscles moving in a specific way. The brains of children diagnosed with ASD don't cue into social information the way others do.[18] These social missteps seem to be tied to an inability to associate others' actions with their own. Whether or not this is really about the mirror system still remains to be seen. Regardless, it is clear that our actions form the basis for understanding how others act.

Charles Darwin defined an attitude as a collection of movements, such as a specific posture, that depicts how a person is feeling at a particular time. Sir Francis Galton also talked about attitudes as bodily inclinations. William James believed that the basis of emotions is the bodily experience of emotional states. Our body not only plays a major role in our ability to feel emotions, it also affects how we resonate with the feelings and intentions of the people around us.

CHAPTER 8

The Roots of Social Warmth

In 1957 at the University of Wisconsin Primate Laboratory run by the psychologist Harry Harlow, Jane, a tiny one-day-old rhesus monkey, was separated from her mother. In the wild, this sort of separation would mean almost certain death for the little monkey, but Jane would be taken care of by experienced animal laboratory technicians and be well-fed, warm, and clean. Jane was placed alone in a wire cage so that Harlow and his research team could study the nature of love.

The first signs of love and affection in humans are those between infants and their mother. Much of our ability to emotionally connect and empathize with others is thought to arise from this intimate connection. But what drives the love of an infant for her mother? How does this initial love for our mother translate into our ability as adults to show affection for a lover or a spouse?

In the 1940s and 1950s, when psychology was dominated by theories from psychoanalysis and behaviorism, the conventional wisdom was that the strong attachment between mother and infant was driven mostly by an infant's most basic need: the need for food, primarily breast milk. Infants were thought to associate their mother with the reduction of hunger, and any feelings of love and affection toward the mother were considered byproducts of this association. Harlow wasn't convinced of this view. He knew, thanks to Pavlov's experiments with dogs, that almost anything can become positively associated with food. Every time Pavlov gave his dogs a steak, he would ring a bell. After a while, the dogs began to salivate at the sound of the bell, even when a steak was no longer part of the deal. Importantly, after an even longer while, the bell would stop triggering this salivation effect, and the link between the bell and the meat disappeared. This kind of association seems very different from the love between a mother and child. Even when our mother is no longer our primary provider, human affection doesn't usually wane. If anything, it strengthens into a lifelong bond. This sort of affection is difficult to explain by the simple satisfaction of basic needs. Harlow wondered if affection in itself was important for healthy development, no less vital than food or water.

Harlow's ideas were in stark contrast to a popular view of the time, that affection served no real purpose for human development. Parents were often warned that too much affection could lead to psychological issues, not help curb them. "When you are tempted to pet your child, remember that mother love is a dangerous instrument," wrote John Watson, a leading psychologist of the day.[1]

Harlow was initially hired by the University of Wisconsin to study the conditions under which rats learn to navigate through mazes to get food, but the university was taking a long time giving him the space he needed to get his rodent work done. After hearing him complain over a dinner party about his nonexistent laboratory, one of his friends suggested he start working with monkeys instead. So Harlow turned a vacant building down the street from the university into state-of-the-art housing for a monkey colony. He had a hard time testing adult monkeys in single cages and found it was easier to work with infants instead. The infants had to be kept in incubators and then in small cages with a piece of cloth diaper on the bottom to absorb the waste.

In contrast to the wire siding, which was hard and cold, the soft cloth diaper attracted the baby monkeys. When the cloths needed to be changed, it wasn't uncommon for the little monkeys to cling to them and throw temper tantrums, similar to the behavior of human children who won't go anywhere without a favorite soft blanket or stuffed animal. What was this attachment to the cloth all about? The diapers certainly weren't fulfilling any of the infants' basic needs, like water and food.

Harlow thought that the infants might derive psychological comfort from contact with the warm, soft, and furry cloth because it had some of the characteristics of their mother's body. To test the idea that what he called "contact comfort" was important to the baby monkeys, Harlow conducted an ingenious experiment in which the infants were paired with two different types of surrogate mothers. One was made from a block of wood, covered with spongy rubber and soft cotton terrycloth. This pretend mother also had a 100-watt lightbulb inside that radiated heat. There was even a round piece of wood at the top with marks for two eyes and a nose. The result was something soft and warm. A second surrogate was built out of wire mesh, but without a terry cloth cover. It was hard to cuddle with this second mother, who amounted to not much more than a wire shell.

Each surrogate mother was placed in a different cubicle and both cubicles were attached to the infant monkey's living quarters so that the infant monkey could easily go back and forth between the mothers. For some of the monkeys, a bottle of milk was attached to the wire surrogate; for other monkeys, the cloth surrogate had the bottle. Strikingly, Jane and the other baby monkeys who were tested chose to spend most of their time clinging to the cloth surrogate mother, regardless of where the milk bottle was. If the wire surrogate had the milk, the baby would drink as much milk as possible as quickly as possible, and then run back to the cage with the cloth surrogate. When a scary new toy, a mechanical teddy bear beating a drum, was put near the infant, the infant would run to the cloth mother, regardless of whether she provided the milk. It seemed that psychological comfort associated with close contact was a driving force in the development of a monkey baby's attachment.

In other studies, Harlow raised one group of babies with a warm cloth surrogate mother and another group with a cold wire surrogate; both had a milk bottle attached. Even though the monkeys gained weight at the same rate, those with wire mothers more often had diarrhea and digestion problems. Physical discomfort, especially digestion issues, are often a sign of psychological stress; thus a lack of physical contact comfort seemed to be psychologically stressful for the monkeys.[2]

We naturally assume that our basic biological needs trump everything else, but Harlow made a striking assertion: the wire mothers who provided milk were "biologically adequate but psychologically inept." His work is still frequently cited as a prime demonstration of how important close contact between mother and child is in building a child's healthy psychological disposition. The cloth surrogate was preferred by the baby monkeys because she was furry, soft, and warm, like a real monkey. Being raised by a warm surrogate likely served as a substitute for the missing social warmth of a real mother. Our brain doesn't always separate the physical from the psychological.

Warm Feelings or a Cold Shoulder?

The connection between temperature and social comfort is apparent from birth. Our caretakers provide love and support and hold us close. Through these intimate moments, we learn to link warm temperature with being in close proximity to others. This association also occurs later in life. When lots of people are in a room together, be it on an airplane, in a classroom, or in an elevator, ambient temperature increases due to the emission of body heat. Warm weather in general is also associated with more close interactions, although not always in a positive way. Crimes that involve interpersonal contact, such as assault, occur more often during hot spells.

Our language illustrates the connection between physical and social warmth. For instance, we describe our friends as "warm and sweet" and our foes as "cold-hearted and hard." A relationship may be "warm and loving," or you may have been given "the cold shoulder." These metaphors arise because we understand our emotions by analogy to the physical world. We even activate some of the same brain states when we think about social warmth and when we physically experience warm temperatures. One important consequence of this connection is that our physical sensation of warmth or coldness has the power to influence our judgments and behaviors—and we often don't realize that this is happening.

Consider an experiment in which neuroscientists invited volunteers to have their brain scanned while they underwent a series of activities.[3] First they read loving messages from close friends and family, such as "Whenever I am completely lost, you are the person I turn to" and "I love you more than anything in the world." In the second part of the experiment, the volunteers sometimes held a warm pack and other times squeezed a rubber ball. Volunteers reported feeling warmer when they read the loving messages than when they read neutral ones like "You have curly hair" and "I have known you for ten years." They also reported feeling more socially connected when they held the warm pack than when they squeezed the ball.

There is growing evidence that humans are born with the capability to make the association between warmth and well-being, trust and safety; in other words, this capability may be hardwired in our brain. The bit of brain tissue in question, the insula, is folded deep inside the brain. It is thought to be involved in the processing of both physical temperature and social temperature, namely, trust, empathy, social exclusion, and embarrassment.

Insula is the Latin word for "island." When you peel off the outer layer of the brain, you find a portion of the cortex that does look somewhat like an island with hilly terrain. The insula registers both physical and psychological experiences, helping to make the crossover between temperature and social connectedness seamless. The insula functions as a communication hub, a relay between the physical and the mental and back again.

This association of physical and social warmth also extends to our actions. Researchers have conducted experiments asking people to rate hot and cold therapeutic pads under the guise of participating in a study on product evaluation. They're asked how good the pads are at generating warmth or coldness and whether they would recommend the product to their friends or family. Then they're asked to play an online investment game in which they have to decide how much money to invest in a trustee, with the hope of receiving considerable gains on their investment. After touching a cold pack, players are less trusting of their anonymous partner, investing less money, than when they touch a hot pack. Researchers found that the insula was more active when people experienced cold temperatures and when they made trust decisions after experiencing cold, suggesting overlap in the brain areas used to gauge temperature and levy decisions about how much to trust someone.[4] Having a cold physical experience makes us less likely to act in trusting ways.

These findings certainly make you wonder about how temperature might affect all types of decisions. In the courtroom, for example, are judges more likely to be lenient in their decisions if the room is warm? Is the stereotype of the warm, friendly Italian compared to the cold Swede due, in part, to the average ambient temperature in which these groups live?

The makeup of our brain tells us that the link between physical and social warmth shouldn't operate in just one direction. If our neural temperature gauge works double-time to understand social interactions, then not only should physical warmth beget social comfort, but the opposite should also hold true. In one experiment, when people were asked to remember being rejected by a former boyfriend or girlfriend, they reported that the room felt colder than when they remembered a socially inclusive experience. When we feel rejected or isolated, we also tend to be more interested in warm food (a bowl of hot soup) and hot drinks (a nice cup of tea).[5]

Our quest for physical warmth when we feel socially spurned might help explain an interesting experience my husband and I had with our daughter. As any parent knows, the first time you leave your child overnight with another caregiver makes you extremely anxious. The possibility of a sound night's sleep and the chance to linger in the shower in peace makes the separation a little easier for the parents, but the entire ordeal is difficult, especially if your kid can speak and is quite vocal about the fact that she doesn't want you to leave her.

The first time my husband and I left our daughter, Sarah, was when she was nearly two years old. We had each been on several business trips for nights out of the house, but never both at the same time. After nearly two years of traveling separately, we figured it was time to get away together. The plan was to leave Sarah with her grandmother, "Munga," as Sarah called her, and get away to the woods for a long weekend alone.

My husband and I found a lovely little bed-and-breakfast near the Point Reyes National Seashore, an out-of-the-way place with limited cell phone reception and a landline only for emergencies. We slept, ate, hiked, and slept some more and returned from the technological void rested, relaxed, and eager to hear how things had gone.

Not surprisingly, my mother reported that there had been tears when we left but they had stopped pretty soon after our car pulled out of the driveway. Once Sarah had come to terms with the fact that we had really left, however, she didn't want to watch a movie or read books or play with the new train set my mother had bought her. She wanted to wear her comfy, fuzzy, and warm pj's. And she didn't want to take them off. Her grandmother thought it was a bit odd but was happy to oblige.

At the time, I wasn't really sure what to make of my daughter's newfound obsession with her pj's. And when it didn't last (once my husband and I were back, she wasn't all that excited about them anymore), I didn't give it a second thought. Until, that is, a few months after our trip, when I came across Harry Harlow's classic psychological research connecting physical warmth with feelings of love and closeness. I realized then that there might be a connection between Sarah's desire to stay in her pj's and her need to feel safe, warm, and taken care of.[6] Harlow wasn't dressing his monkeys in comfy pj's, but what was happening wasn't so far off. Physical warming may make social isolation feel less severe.[7]

Loneliness really seems like a social coldness. This surely sheds light on the self-help books *Chicken Soup for the Soul.* For close to twenty years, this series has captured the hearts of millions of readers around the world by telling real-life stories of success and love. People know to turn to these books to find inspiration after breakups or in times of social isolation. What many people likely don't know, however, is that having some warm chicken soup while reading these books may also not be a bad idea. Actually, there are a whole host of self-medicating activities that may be advisable when we feel lonely—ones that can be easily incorporated into our lives when we're aware of the mind-body connection. Vacationing in warm locales, putting on a cozy sweater, or even having a hot toddy may contribute to feelings of being loved and included. The opposite seems to be true as well. In cold winter months, you are more likely to opt to watch a feel-good romance.[8] We seek emotional warmth in the form of romantic chick flicks. Our body and the surrounding temperature have a profound effect on our mind.

This interchangeability of feelings and temperature also helps us understand psychological disorders like seasonal affective disorder, otherwise known by the fitting acronym SAD. People with SAD experience symptoms of depression during the winter months, when it's dark outside even during the day. This disorder is different from ordinary depression because folks with SAD are otherwise healthy, especially in the sunnier summer months. Research on SAD has predominantly focused on the connection between reduced daylight and depression, but colder temperatures might contribute to an increase in patients' sadness and loneliness. Cold temperatures during the winter may also magnify their feelings of depression. Though UV sun lamps are often prescribed for people who suffer from winter depression, they may also benefit from the power of warmth.

Simply put, being warm can make people feel better and more connected. Perhaps it's no accident that some of the most important political meetings in history have taken place in warm, intimate environments. Camp David, for example, is tucked in the wooded hills of Maryland. U.S. presidents since Franklin D. Roosevelt have brought world leaders together at Camp David to navigate treacherous political issues and broker important deals. Jimmy Carter oversaw the Camp David Accords, a landmark peace settlement in the Middle East in 1978 between President Anwar al-Sadat of Egypt and Prime Minister Menachem Begin of Israel. President Obama held the 2012 G8 Summit at Camp David. Sitting by a warm fire likely helps foster feelings of inclusion, of being of one mind. The end result is a path to mutual understanding and decision making. Warmth makes people feel socially close and connected.

Rejection Hurts, Physically

For decades, neuroscientists have been keenly aware that a specific brain circuit is involved in registering physical pain. Whether you get pricked with a needle, burn your hand, or sprain your ankle, many of the same neural circuits come alive to process the pain. This "pain matrix" includes brain areas such as the insula, the cingulate cortex, and the somatosensory cortex, which registers information coming in from our senses. Scientists have discovered that, just as in the connection between coldness and loneliness, some of the same bits of neural tissue involved in recognizing our physical pain also give rise to painful feelings and emotions.[9] We understand psychologically adverse situations, whether it's "hurt feelings" or a "broken heart," as physically adverse ones.

Using the same brain systems to register social and physical pain makes evolutionary and economical sense. Rather than developing an entirely new brain area to register social pain, we evolved so that our more ancient pain systems perform the same functions. Perhaps the Stanford University neuroscientist Robert Sapolsky said it best: "Evolution is a tinkerer and not an inventor."[10] We deal with social pain the best way we know how: we feel it physically.

It's easy to see how our physical pain system could have evolved to register social pain. Many primates, particularly humans, have a long infancy, which means that maintaining social connections at an early age (for food, shelter, and protection) is critical for survival. If being separated from a caretaker is a threat to survival, feeling emotionally hurt by this separation could offer an adaptive edge, helping to keep caretakers close at hand. Perhaps those infants who were best at using their pain system as a social alarm when they felt distant from their caretakers were the ones who thrived, resulting in the evolution of a system that serves two purposes.[11]

Two UCLA neuroscientists, Naomi Eisenberger and Matt Lieberman, support the view that our physical and social pain systems are one and the same. It all started in 2003, when they did an experiment in which they asked volunteers to take part in a computer game known as Cyberball.[12] Cyberball appears to be a virtual game of catch with two other players whose computer is networked to the volunteer's. The volunteer can't see the other people playing; he's told something about them—their names, ages, and a little information about their interests and backgrounds. For a while, the three play catch, but at some point the other two players stop including the volunteer, tossing the ball back and forth only with each other. He can only sit and watch as he's excluded, shunned from the game.

In reality, there aren't any other players; the game is controlled by a computer. But the volunteers don't know this. While the volunteer played and then was excluded from the game, the scientists peered inside his brain and discovered that part of the neural pain matrix—specifically the insula and the anterior cingulate cortex (ACC)—came alive. Along with its role in processing negative emotions, the ACC acts as a neural alarm system, detecting when an action, response, or event is in conflict with a larger goal. For example, the ACC might become active in a husband when his wife asks if the outfit she is wearing makes her look fat, and, without thinking, he responds, "Yes." When we make a social error like this, the ACC generates a unique electrical signal that floods the rest of the brain with information that there is a problem. Because of this, the ACC is often talked about as our "oh, shit" sensor, giving rise to the realization that something is wrong. It's not surprising, then, that the ACC becomes active when we are rejected or in an adverse social situation. Physical pain, the most basic signal that there is a problem, also activates this brain area.

Because the brain doesn't always make a clear distinction between physical and social pain, some of the ways we go about alleviating physical pain can help lessen social pain too. When people take acetaminophen (Tylenol) over the course of several weeks, they report less daily social pain, and their brain's pain matrix is also less reactive to social rejection. A daily dose of Tylenol diminishes the hurt feelings that often accompany being socially rejected likely because it reduces the sensitivity of our neural circuits involved in pain.[13]

Social exclusion is a normal part of life. We have all, at one time or another, felt disliked at work, spurned by a partner, or snubbed by friends. Even though it's unpleasant, social rejection seems pretty different from a physical injury. Yet these experiences share a common biological substrate in the brain. Evolution's solution to our need for caretaking seems to be instilling in us a need for social connection and a sense of distress when those connections are severed. It really does *hurt* to lose someone we love.

Understanding the link between the mental and the physical also arms us with clues for how best to interact with others—especially when we need them to perform well. At work, for instance, calling out individual workers for failed projects or a lack of collegiality may trigger a cascade of neural responses in a colleague's pain matrix, responses that result in less productivity and worse future performance rather than better. When our social alarm systems are triggered, we have less brainpower to think productively about the task at hand. Instead, fostering relationships that help teams of people feel connected might do more to boost work performance. When we feel connected, we work better. Team-building exercises that encourage groups of people who work together to feel more physically trusting may be just what people need in order to be mentally connected too. In the classic trust exercise, you stand with your back to a group of people and fall backward into what you hope will be their comforting and supportive arms. Our mental and physical worlds cannot be carved up into neat, separate boxes. Once you understand this, you may be more likely to guard against emotional distress with the same care you take to ensure that no physical harm befalls you or those around you.[14]

Distance Matters

In a follow-up experiment in Henry Harlow's Primate Laboratory, Jane and the other baby rhesus monkeys were moved from their cages to a new room littered with unfamiliar toys that produced odd sounds and jolting movements. A mechanical teddy bear that played a drum was the most unsettling. Scared to be in a new place, Jane sought out her warm and soft cloth surrogate mother. Like any young child who finds herself in a strange and unfamiliar situation, Jane made a beeline for her mom, to which she clung until she felt brave enough to explore her surroundings, going back every so often to make sure her mother was right where she left her. Some of the other monkeys had only their cold, wire surrogate mothers to keep them company. These monkeys behaved very differently, traveling randomly around the room, almost as if they were looking for their real mother. Though the wire mother had milk, the young monkeys didn't cling to her.

Not only is warmth an inherent part of social contact with a caregiver, but being close matters too. There is value in infants staying within arm's reach of a parent so they can be protected from predators or foes. Physical closeness equals connection and safety. What's striking is that the baby monkeys in Harlow's lab didn't seem to feel there was any value in staying close to their wire mother, even though she had food. Could this be because an emotional distance is largely interchangeable with physical distance? If the baby monkeys didn't feel socially connected to their wire mother, they may not have felt the need to be physically close to her either.

The most famous demonstration of the strong tie between physical and emotional distance comes from a set of studies conducted by the Yale University psychologist Stanley Milgram in the 1960s. Sparked by an interest in how people could condone atrocities such as the Holocaust, Milgram set out to see just how far volunteers would go to follow the instructions of an authority figure. What he found was that even when obeying authority meant doing something that went against most people's conscience, such as giving a complete stranger an electric shock, most volunteers did as they were told. Milgram also found that how physically far away two people were from each other made a big difference in terms of the likelihood of administering shocks. A *physical* closeness seems to beget a *psychological* connection.

Here is the basic design of the Milgram experiments: People were offered $4.50 to be part of an experiment on "human memory"—$4.00 for the experiment and 50 cents for carfare ($4.00 was a pretty good wage for an hour's worth of work in the 1960s, especially if you were a starving college student, and 50 cents more than covered round-trip bus fare). When the volunteers arrived at Milgram's lab, they met the experimenter, a stern man who introduced himself as Mr. Williams and was dressed in a gray lab coat. Mr. Williams introduced the participant to another person who appeared to be volunteering for the same study but was actually an actor hired to play the part. Then Mr. Williams announced that one of the volunteers was going to play the role of the teacher and the other was going to be the learner. To determine who was going to assume which role, both drew a slip of paper from a bowl. The slips were rigged: they all said "teacher." The second volunteer pretended that he had drawn the learner role.

"Teacher" and "learner" were put in different rooms connected by an intercom system, and the teacher was given a list of word pairs the learner was to memorize. The teacher was instructed to read the list over the intercom, then, starting at the beginning, to reread the first word in the pair. The learner was to respond by picking the word that satisfied the pair. Every time the learner got a word wrong, the teacher was to give the learner an electric shock, increasing the voltage with each wrong answer. So that the teacher knew what the shock felt like, the experimenter shocked the teacher before the experiment began. The teacher really believed that he was administering a shock to the learner. Indeed a tape recorder yoked to the shock machine played prerecorded cries of agony when the teacher administered a shock for a wrong answer.

Milgram polled some Yale students (who didn't take part in his experiment) and psychiatrists beforehand to find out how many people they thought would actually follow the experimenter's instructions to shock a stranger. Both students and psychiatrists thought very few people (maybe 1 percent) would do so, but Milgram found that 65 percent of the volunteers, twenty-six out of forty, continued shocking until they had given the maximum 450-volt shock three times. A standard U.S. socket maxes out at 120 volts, so 450 volts hurts a lot.

Both men and women shocked the learner at the same rate, and compliance was also strikingly similar whether Milgram conducted his experiment with Yale students or a broader sample of the general population. One factor that did affect how likely someone was to give a stranger a shock, however, was the physical distance between the two people. The teacher was less likely to shock the learner if the learner was physically closer, in the same room rather than a separate room. Being physically close seemed to relay a sort of mental closeness that minimized the suffering the teachers were willing to inflict.[15]

Why does physical distance alter our willingness to impose harm on someone? We understand others, and relate to others, in part by how close we are to them physically. Physical distance information is actually built into the design of the human brain. The computation of information about how close we are to a looming threat shifts from the frontal cortex to more rudimentary, pain-related regions toward the middle of the brain as we get nearer to a possible danger.[16] When we are in close physical proximity to someone or something, our brain's more primitive emotional regions perk up, which could help us better understand what others are feeling. A close physical distance paves the way for a strong emotional connection, while greater distance capitalizes on the association we have between distance and disconnection.

The effects of physical distance on feelings of psychological closeness offer an important lesson for our interactions in the virtual world. These days, in-person meetings are regularly traded for video conferencing, whether to pitch to clients, strategize among board members, or even interview for a job. Although virtual interactions have some advantages, limiting the cost and time of travel, they may have downsides. Our mind cannot be separated from our body, and, though we may wish otherwise, research shows that physical distance encourages the perception of being psychologically distant. In short, if you want to see eye to eye with someone, there may be value in being in the same room with him. If we use information about physical distance to render judgments about psychological distance, it may be harder to get on the same page with others in a negotiation, to trust them, and to come to a mutually beneficial outcome in the end. If you are a job candidate and have the choice of an in-person interview or one over a virtual connection, you will likely benefit from choosing the former. Our physical environment activates feelings of closeness or remoteness without our even knowing it.

Our body shapes our attachment to others. Just as loneliness is partly built on physical coldness and social pain on our physical pain system, moral transgressions can also stem from the physical nature of disgust. There is something to the saying "He who has clean hands and a pure heart may ascend the hill of the Lord." There are parallels between physical and moral contaminants.[17]

Cleansing the Body

What do Lady Macbeth, religions throughout history, and parents who wash their kids' mouths out with soap have in common? They all believe in a link between physical and moral cleanliness. Both the Christian ritual of baptism and the Jewish ritual of the mikva involve metaphorical and physical baths. "Arise and be baptized, and wash away your sins," the Bible says. The link between physical and moral cleanliness is reflected in Islamic traditions too. Many washing rituals are believed to help root out psychological demons. In Shakespeare's tragedy Lady Macbeth attempts to cleanse her conscience of her culpability in the murder of King Duncan by washing her hands repeatedly.

People often think about morality in terms of cleanliness. That's why, when we recall an immoral past behavior like cheating on a test or lying to someone, we often have an urge to physically clean ourselves. The physical act of cleaning helps us feel psychologically cleaner. It also tends to be specific. When volunteers in a role-playing game were asked to communicate a mean lie via voicemail or email, they were more likely afterward to prefer products that would help clean the body part they used to do the dirty deed. People who had left the nasty voicemail preferred the mouthwash over the hand sanitizer; those who had sent the nasty email showed the opposite preference.

When we soil our hands, we wash them; when we drink sour milk, we wash out our mouths. This specificity of cleaning is quite functional; it gets rid of the adverse substance, reducing the risk of contamination or disease. The fact that this exactness spills over into our psychological state is a prime example of how the regulation of moral behavior may be built on procedures of physical cleansing and disease reduction. It's good evidence for a reuse of the same neural circuits that helped us stay disease-free. We have evolved to tack on new functions to existing behaviors already in place. In this way, our ability to reason about abstract ideas, from love to morality, is driven by our concrete experiences acting in the world.

Cleansing acts actually make us feel better about ourselves and help restore moral cleanliness. That shower that people take after they have cheated on a spouse may do a lot more good than simply getting rid of the physical evidence; it also helps rid them of a guilty conscience. People routinely deal with guilt and other negative feelings after committing bad deeds by physical cleansing. Sometimes it is as simple as washing one's hands of the situation.[18]

In the biblical story of the trial of Jesus, he was arrested and brought to the governor of the province, Pontius Pilate, who was reluctant to sentence him to death. Pilate publicly washed his hands, telling the crowd that he would not take the blame for Jesus' death and was washing his hands of Jesus' blood.

The body embodies abstractions physically. An ethical violation is a soiling of your character. To feel better, clean your body. If you already feel physically clean, but you have witnessed someone else doing something immoral, you may make harsher moral judgments on issues ranging from abortion to drug use. Dov Cohen, a psychologist at the University of Illinois, has been studying this correlation between the physical and the moral for the past several years.[19] One question Cohen asks is how universal the link is between physical and moral disgust. He has capitalized on some interesting differences between cultures as a way to answer that question. There are some real differences in how religions think about moral transgressions. For Muslims and Protestants, just having bad thoughts is considered impure. But for Hindus and Jews, it's the actions that matter; you can think all the negative or immoral thoughts you want as long as you don't act on them.

By comparing how people with different religious backgrounds rated how morally wrong certain thoughts and behaviors were, Cohen was able to document the power of cleanliness. He discovered that, when people rubbed their hands together as if they were washing them, they made more severe moral judgments of others than when they weren't mimicking washing movements. The exercise was presented to volunteers as a way to warm up their hands for a video game they were going to play; they were not conscious of making washing movements. Most striking, Cohen discovered that the hand-washing movements produced greater condemnation for immoral beliefs among Muslims and to a lesser degree Protestants, but for Hindus and Jews, only the disgusting behaviors were judged morally wrong. Even though the link between physical and psychological cleanliness seems to be universal, what counts as impure varies across cultures.

When we do something bad, washing ourselves makes us feel better. But when we bring our own feelings of disgust to bear in making moral judgments about others, being clean makes us harsher about other people's "dirty" behaviors—and for some religions, other people's dirty beliefs too.

The fusing of mental and physical cleanliness isn't limited to moral transgressions or bad deeds done. As echoed in a line in one of Oscar Hammerstein's most famous songs, "Gonna wash that man right outta my hair," we believe that we can wash away our sins (or at least our bad feelings about a situation or relationship). Yet we also believe it's possible to wash away the good, specifically good luck, as if washing our hands is akin to "wiping the slate clean." It's not uncommon for an athlete to go an entire season without washing an armband or a pair of socks when they are on a winning streak because they fear physical cleansing will get rid of their psychological advantage.

It may not be particularly surprising that how much people are willing to bet when they gamble is linked to whether they have previously been on a winning or a losing streak. But it's also the case that people's wagers depend on whether they have just washed their hands. When gamblers don't wash, they bet more money on the next round if they were on a winning streak than if they were on a losing streak. Among those who did wash, having a winning or losing streak had no impact on how much they gambled. People are driven by the idea—at least on an unconscious level—that washing can remove the influence of their past gambling streak (whether good or bad). So when people wash, the gambling outcomes of the past no longer seem to matter.[20] When you understand that the brain views physical and mental cleansing as largely interchangeable, all sorts of rituals, behaviors, and decisions that once seemed to have no rhyme or reason make a lot more sense.

Not only can cleaning our body enhance psychological well-being, moving our body does too. In the next chapter we explore the relation between exercise and sharp thinking. Most books on exercise tout the importance of exercise for physical health. These books, as a whole, though, focus less on the story of how exercise can be used to be sharper mentally.[21] Exercise is a tool to increase mental fitness.

CHAPTER 9

Movement

HOW EXERCISE ENHANCES BODY AND MIND

The Running Mind

I wear a Fitbit on my left wrist day and night, while washing dishes and even in the shower. In fact the only time I take it off is when I get an email message warning me the battery is low. And then I remove it reluctantly, preferably before I go to sleep and am not likely to be walking around. Fitbit is a fitness tracker I wear as a wristband that tracks my steps and the minutes throughout the day when I am active. My friend Melissa turned me on to it after she got one as a way to make sure she was keeping active during her pregnancy. We could compete against each other, she said. The one who takes the most steps a day wins. Never one to turn away from a little competition, I jumped at the chance, not only to try to show up my friend but because I was in the middle of teaching a class on neuroscience and education at the University of Chicago and we had just finished reading several new research papers on the power of exercise in changing the brain. If tracking my activity with the Fitbit motivated me to take more steps throughout the day, then why not try it? It seemed like a good idea to practice what I was teaching.

I am a runner and make sure to get in several runs a week, but before I had the Fitbit I never realized how much of a couch potato I was when I wasn't lacing up my sneakers. The device made me mindful that the little things I did could add up to a significant amount of mileage a day. Just parking my car at the far end of the parking lot when I made a trip to the grocery store or taking the stairs rather than the elevator to my office on the third floor of the psychology building made the difference between getting in eight thousand steps a day and exceeding my goal of ten thousand.

The Fitbit is also a fun conversation starter. Because wearing the gadget signals that you are buying into the power of movement, it's easy to strike up a conversation with strangers at restaurants and on the train who are wearing a similar device. Of course, most of the conversations I find myself engaged in with fellow Fitbitters have to do with how important exercise is for maintaining a healthy body. It's rare that folks think about exercise in terms of benefiting the mind. Yet fitness contributes to both physical and mental health. Brains look and function differently according to whether they are housed in inactive or active bodies.

Exercise simulates the creation of new brain cells, a process known as neurogenesis.[1] Some of the first research documenting the link between physical activity and the brain was done using mice. Mice raised in "enriched environments," in which they had toys, exercise wheels, and lots of opportunities for social interaction, grew more new brain cells than their litter mates housed in standard laboratory cages. Scientists weren't actually sure which part of the mice's surroundings caused the new cell growth, so, in the late 1990s, researchers at the Salk Institute at the University of California, San Diego, conducted studies to find out. They systematically examined the different parts of the mice's environment in order to unveil what exactly was responsible for the neurogenesis.[2]

The experiment that conclusively demonstrated the striking power of exercise on brain function followed a simple protocol. The scientists began by giving all of their young mice a chemical that could track the brain cells when they divided and new cells were produced. They then offered some mice access to "exercise equipment"—a running wheel that they could use as often as they liked. Another group of mice didn't have the opportunity to exercise and led a fairly inactive lifestyle. After several weeks, the scientists sacrificed the rodents so they could see if and how the brains of the two groups differed. They found striking differences: those mice that had been on the move had more new brain cells, roughly twice as many, compared to their sedentary counterparts.

To make sure it was the vigorous exercise that had changed the mice's brains, the Salk researchers enrolled another group of mice in their study. The mice in this new group learned how to navigate a maze. They had a lot of mental exertion but not as much physical activity as their wheel-running counterparts. Surprisingly, the opportunity to expend cognitive effort didn't lead to nearly as much proliferation of new brain cells as running did. The lesson here? Even though we are often tired after a long day of work and it feels as though we have run a marathon, it's not the same as actually hitting the track—at least to our brains. Vigorous exercise is important for growing new brain cells.

The mobile mice showed the most new cell growth in a

seahorse-shaped area deep inside the head, the hippocampus. This is one of the major brain centers that helps mice and men turn what we learn into long-lasting memories.

Fit Children

The brain has an impressive plasticity, particularly in childhood, and physical activity can improve kids' mental functioning. Charles Hillman, a professor at the University of Illinois, has devoted much of his research career to documenting the power of exercise in altering kids' brainpower. His work clearly shows that the time spent on physical activity does not come at the expense of academic achievement; rather fitness enhances what kids can do in the classroom.

In a recent study, Hillman and his colleague Art Kramer and their research team pulled together data on the physical fitness of a group of nine- and ten-year-olds. They scanned the children's brains as the kids completed a series of cognitive challenges designed to test their thinking, reasoning, and memory skills and found that the fittest kids performed best on many of the memory tests. Even more telling, children's physical fitness level roughly corresponded to the size of their hippocampus. Just like the wheel-running mice, the fittest kids had the most developed hippocampus.[3]

To further validate the link between a fit body and a fit mind, Hillman and Kramer also tested whether they could actually find direct benefits from a session of exercise on young kids' brain functioning.[4] The researchers asked a group of children to visit their laboratory on two separate occasions. During one visit, the children took part in a short bout of exercise: twenty minutes of walking at a fairly vigorous pace on a treadmill. On the second visit, the kids rested, sitting quietly in a chair, for twenty minutes. On each visit, after the kids had either rested or exercised (and the heart rates of the kids who had exercised had returned to normal), they were given a series of cognitive challenges. In one challenge, they were told to focus on one critical piece of information presented on a computer screen and ignore anything else that popped up. This mental activity is not unlike a situation a child might face when doing homework and the cell phone pings with a text message from a friend. To successfully complete the schoolwork, the child must focus on the academic material and ignore the tempting distraction. It's also similar to concentrating on an exam and not letting your mind wander to thoughts of what you are going to do with your friends after school. The mental challenges in the experiment, in other words, mimicked the focus that kids need to maintain in order to succeed in school.

Not only did the children perform better on the cognitive tests when they exercised rather than rested, but their brain functioned more fluently after exercise. Neural activity emanating from frontal and parietal brain areas, activity known to reflect our ability to control our attention (something of dire importance in school), was enhanced after the kids had exercised compared to when they had been sedentary.

For a long period in human evolution, our ancestors lived as hunter-gatherers. Moving across plains and mountains to hunt game and gather nuts and berries was necessary to our survival. This means that our minds and bodies evolved in the setting of an active lifestyle. Physical activity seems to be programmed into our genes.[5] But the amount of activity that young kids, adults, and senior citizens get today is usually well below what we are genetically predisposed to do. The consequences of a sedentary existence are evidenced by ill health in body and mind. Children who are more physically fit perform better on academic tests. Elderly people who are active have a lower risk and incidence of memory loss and loss of other important cognitive functions. Providing kids with opportunities to be active and to exercise helps hone their mental as well as their physical muscles. And a regular exercise regimen for adults helps prevent mental decline.

As testing becomes an increasingly larger part of our academic culture and school budgets continue to be tightened, recess, gym class, and physical activity are targeted for elimination in the mistaken belief that getting kids to spend more time in the classroom is a cheaper and better way to boost test scores than giving them breaks to run around. Yet discoveries in brain science tell a different story. If we want to produce kids with the most brain health, ability to focus their attention, and superior thinking and reasoning skills, then we need to add a fourth "r," recess, to the curriculum. We also need to make sure that kids are getting the opportunity to exercise outside of school, as it's not atypical for a child to have no physical activity outside of school hours. Knowing that the fitness of the body has a big impact on the fitness of the mind provides us with a clear prescription for children: get them moving.

Adults

Physical activity is also key to a better brain after adolescence, although we know less about how fitness relates to brain function in young adults. Our cognitive functioning is at its peak in our late teens through our thirties. Of course, even at the peak of our mental capacities, we don't always perform at our best. We have all experienced situations where the stress of an important test, speech, or job interview robs us of the brainpower we would normally have to perform well. Exercise can help us get access to all our cognitive resources.

Short bouts of exercise specifically benefit the functioning of a network of brain regions that include the prefrontal cortex, the parietal cortex, and the hippocampus, which support thinking and reasoning and especially working memory. You can think of working memory as a kind of mental scratch pad that allows you to work with whatever information is in your consciousness. It helps you focus on what is immediately relevant to a task and to screen out what is irrelevant. Working memory is one of the major building blocks of IQ.[6]

An important detail about working memory is that it is limited. We have only so much of this brain resource at our disposal. In pressure-filled situations—taking a test, pitching a client, or interviewing for a job—we have even less. Stress drains our working memory, but exercise jump-starts the brain regions that support working memory, so it improves thinking, elevates mood, and reduces stress. And this exercise boost tends to be biggest for those people who come to the table with lower amounts of working memory to begin with.

Some people naturally wield more brainpower, more working memory than others. One way for those with less working memory to perform as if they have more is by taking part in a short bout of exercise. Ben Sibley and I discovered the advantage of exercise for those with less working memory several years ago, when we were both on the faculty at Miami of Ohio University. Ben had recently found that a short bout of exercise had an immediate and positive impact on people's ability to focus their attention. Since focusing on some information while keeping irrelevant information out of consciousness is at the heart of working memory, we wondered if the benefits of exercise might be most extreme for those people who had the most trouble focusing to begin with.[7]

We began by inviting about fifty undergraduates to Ben's lab in the basement of the kinesiology building at the university. We first asked them to take a number of tests that gauged their working memory. An important point about measuring working memory is that it doesn't much matter what people are holding in mind; what's important is that we can measure their ability to focus on some bit of information when they are distracted.

In one task, called the Operation Span Task,[8] participants were asked to solve aloud a math problem that appeared on a computer and was followed by a word:

Is (10 ÷ 2) − 3 = 2? SEA

Is (10 ÷ 10) − 1 = 2? CLASS

Is (5 × 2) − 2 = 8? PAINT

Is (4 × 1) − 1 = 3? CLOUD

Is (6 ÷ 3) + 3 = 5? PIPE

After reading aloud and solving each math problem, the students were told to read the word aloud and remember it. Then the math problem and the word disappeared from the screen. Deciding whether the math problem is correct was not the main goal of this task; we wanted to know how good they were at remembering the words at the end. After a number of math-word pairs (usually between three and five), we asked the students to recall all the words in the order in which they appeared. Even though the students knew they were going to have to recall the words, they didn't know *when* the recall task was coming, so they had to keep the words in mind while they were doing the math problems. Holding information in memory while doing something else is really what working memory is all about.

After getting measures of working memory, we asked everyone to run for thirty minutes on a treadmill set up in the lab. The running was self-paced, but we asked everyone to try to work out at about 60 to 80 percent of what they would consider their max. After that, we asked them immediately to take part in some more working memory tasks, but we replaced the math problems and words with new ones so they couldn't use what they remembered from the first tests to help them out on the second.

What we found was that those folks who came to the table with less working memory to begin with benefited most from a short but moderate bout of exercise. This finding was exciting because, it's not just that adults differ in their working memory from each other, working memory is also something that changes across the lifespan. Young children tend to have less working memory because brain areas such as the prefrontal cortex, which support our ability to hone our attention, are still developing. Working memory also tends to decline in senior citizens. This means that exercise programs that target younger and older people might be especially beneficial, boosting the power of developing and waning working memories.

Being active and working up a sweat can even help you negotiate better. In a study recently conducted at MIT's Sloan School of Management, researchers found that people negotiated better in a deal for a used car or even a compensation package for a new job when they got their heart rate up by walking at a quick pace on a treadmill.[9] But there was a catch. Exercising led to better negotiating only when folks went into the negotiations from the outset confident in their powers of persuasion. For those who came to the table already flustered, pushing up their heart rate with a short bout of exercise led to worse deal making (both in terms of their feelings about the negotiation and how they objectively performed). How we perform has a lot to do with how we interpret our bodily reactions. Confident negotiators viewed their beating heart as a sign they were thriving, but those who dreaded negotiating thought their physiological state was a sign that they were failing, so they performed poorly. Whether we view our racing heart and sweaty palms as a sign of excitement or anxiety has a lot to do with whether we will clutch or choke.

Thankfully, we can all learn to view our ramped-up symptoms in a more positive light. For several years, the psychologist Jeremy Jamieson and his colleagues at the University of Rochester have been exploring the benefits of having people reappraise their higher heart rate or other physical stimulation in potentially stressful situations, such as test taking, public speaking, and having an anxiety-producing social encounter. For instance, Jamieson and his colleagues have shown that getting students to think of a racing heart or sweaty palms as an energy resource can actually help them perform better on tests. Most striking, when provided with tips for rethinking their physical signs, people's views of potentially stressful situations change. If they were nervous to begin with, they still report the situation as demanding, but they now believe that they possess better ability to cope compared with folks who don't get the reframing tips.[10]

Of course, moderation is key in using exercise to boost thinking, reasoning, and negotiating. While some physical activity boosts working memory—especially for those with less to begin with—and may make you better able to cut a favorable deal, more than an hour of intense physical activity before an intellectual challenge is not necessarily advantageous. Long bouts of exercise that lead to dehydration, for example, can deprive the brain of important nutrients you need to function at your best. But exercise can boost your ability to think on the fly and perform optimally, especially in response to stress. When asked directly whether exercise is beneficial before conducting negotiations, one of the MIT researchers involved in the treadmill study cautioned, "I wouldn't suggest doing a marathon."[11]

Long-term fitness is also associated with enhanced thinking and reasoning in young adults. In a recent study that tracked more than a million eighteen-year-old men in the Swedish army, better fitness was related to higher intelligence and more job success. The fitter the soldier, the higher his IQ and the more likely he was to go on to a successful career than his less fit counterparts.[12] Physical fitness is linked to enhanced functioning in important frontal and parietal areas of the brain, which help support working memory and our focus of attention,[13] so it's easy to see how physical fitness could translate into mental fitness. Being fit allows you to recruit more brainpower when high levels of mental effort are required.

Fitness does more than simply give people extra mental computing power. Exercise can also enhance the type of creative smarts for which companies like Apple and Google are known.[14] Not coincidentally, these companies are very into fitness, with in-house gyms and trainers for employees, and they're famous for coming up with new ways to use a product, such as the iPhone and Gmail. Exercise helps the brain see things in new ways. Success in the workplace is not always about putting your nose to the grindstone and pushing through masses of data, paper, and problems. Sometimes it's about knowing when to take a step back so that you can see something from a different perspective, find an unplumbed corner of the market, enhance an old tool in a unique way.

Short bouts of aerobic exercise help a neurotransmitter, dopamine, circulate in our brain. Dopamine plays an important role in many aspects of brain function, such as control of movement, sensitivity, feelings of gratification, and focus of attention. A gradual decline in dopamine generally occurs with aging, but this decline is significantly smaller in exercising animals, even if an animal doesn't begin exercising until later in life. Dopamine also plays an important role in creativity, our ability to think flexibly about a problem from multiple perspectives. An exercise program helps stave off the natural decline in dopamine.[15]

Yet another reason to lace up those sneakers and get out for a walk or run during lunch is that our body and our fitness give us more positive views of the world around us. Compared to their fit counterparts, people who are out of shape and have poor fitness levels judge hills to be steeper. People who experience chronic pain when walking think objects are farther away than folks of the same age who don't have difficulty walking. Older adults who become less fit and less mobile as they age estimate a hallway to be longer than do fit college students.[16] If being unfit leads to judging distances and hills as less traversable, this could lead you to be more sedentary. A vicious cycle emerges, in which an unfit body influences the mind and makes it harder to get moving in the first place.

Later in Life

Before the fitness gurus Richard Simmons and Bob Greene held sway, there was Jack LaLanne, who is often called the father of the modern-day fitness movement. As a teenager, he discovered the power of nutrition and exercise, which gave his life meaning and direction after he had been through a rough childhood. In 1936 in Oakland, California, LaLanne opened what might have been the first health club, complete with a gym, juice bar, and health food store. (He later sold his chain to Bally.) In the 1950s, *The Jack LaLanne Show* first aired locally in northern California and then went national. It still runs today on ESPN Classic. LaLanne preached the benefits of exercise and good nutrition at a time when no one—not even those in the medical profession—paid much attention to the influence of the body on how people felt, thought, and behaved. A walking testament to the impact of exercise on physical health, LaLanne was incredibly fit, with bulging biceps, a wry smile, and an engaging wit. As he famously remarked, "I can't die. It would ruin my image."

LaLanne did eventually pass away, in 2011, at the age of ninety-six. Yet even in his final years he continued to exercise regularly, for up to two hours a day, swimming and lifting weights in his home on California's central coast. In his sixties he made the treacherous swim from Alcatraz to Fisherman's Wharf in San Francisco while pulling a thousand-pound boat. In his seventies he performed another amazing feat in the water, this time swimming over a mile and a half in the Long Beach Harbor while pulling boats that held more than seventy people. LaLanne set an example for aging people, encouraging them to exercise and proclaiming that gyms are not just for the young.[17]

Jack LaLanne may have been one of the first advocates of the power of exercise, but he isn't alone in believing that fitness is important for people of all ages. You can see people in their forties, fifties, and sixties circling the local track or walking the mall in the morning, but you can also find athletes in their seventies, eighties, and nineties. Masters athletic programs can be found around the country. You have to be at least thirty-five to take part in masters competitions, but some of the most exciting competitors are much older. It's estimated that some fifty thousand people worldwide call themselves masters track and field athletes.

Consider Olga Kotelko, a ninety-three-year-old Canadian track athlete, who has a habit of setting world records. It's true that there aren't that many women competing in her age group, but Kotelko is killing her competition. At the World Masters Games in Sydney, Australia, in 2009, her 23.95 seconds time for the 100-meter dash put her squarely with the finalists who were two age brackets younger.

People eighty-five and older are the fastest growing segment of the world's population, so researchers are turning to them for clues to what promotes health and longevity. Most of the work targeting these individuals looks at what foods they eat and their social lives. But in Kotelko's case, scientists are quite interested in how long-term exercise alters the body and the mind.

Unlike LaLanne, who found physical fitness in his teenage years, Kotelko didn't start masters track and field until she was in her late seventies.[18] She grew up on a farm in Saskatchewan, where she was always an active child, feeding chickens and milking cows rather than playing sports. Organized sports weren't readily available, and the ones that were around weren't open to girls. After she retired from her career as a teacher, she got involved in slow pitch softball, and when her softball career was ending in her early seventies, a friend suggested that she might like track and field: it would give her something to do and could be a great way to meet other retired people in her area. She found a coach, and the rest is history—literally—with Kotelko breaking world records at every turn of the track.

Kotelko's case illustrates that exercise prolongs life and promotes health. Scientists have taken samples of her muscle fibers and found that exercise seems to have rolled back time in terms of her cellular health. Usually mitochondria, cell structures that generate energy for cells and muscle, decay in senior citizens, but there is little sign of that in Kotelko. Scientists are extremely interested in why her body doesn't seem to be aging very quickly. They also want to know how exercise prolongs mental health. Striking new research shows that exercise does indeed benefit cognitive functioning in later life. There are clear differences in brain health in fit older adults compared with their more sedentary counterparts, and these differences carry consequences for thinking and reasoning as well as for memory.

Several years ago, scientists averaged the results from roughly two dozen studies in which researchers had randomly selected adults over the age of fifty-five to participate in an exercise training program or to serve as a control group (which did not exercise). Despite all of the differences in the exercise programs that people took part in, when the results were tallied, there was a clear pattern: the exercise groups had greater cardiovascular fitness and mental fitness. Exercise clearly helped improve their working memory too. Older adults who had taken up an exercise regimen performed markedly better in tasks in which they had to focus attention, concentrate, or think quickly.[19]

Carl Cotman, director of the Institute for Brain Aging and Dementia at the University of California, Irvine, and his colleagues want to know how exercise improves brainpower. One biological mechanism that Cotman believes may be at the root of exercise's positive effect on mental health is brain-derived neurotropic factor. BDNF and related growth factors are sometimes loosely referred to as "brain fertilizers" because they help support the survival of existing neurons and the growth of new ones. Rats that run on wheels have increased levels of BDNF in their hippocampus, one of the brain areas important for learning and memory. Not only did these rats show increases in BDNF levels, but the more they had, the better they performed on different sorts of cognitive challenges. Likewise BDNF increases after a short bout of exercise in humans.[20]

Interestingly, carrying a certain copy (called allele) of the gene encoding BDNF, specifically, the methionine-specifying (Met) allele at amino acid 66 of the BDNF gene, is associated with reduced secretion of BDNF and poorer working memory in healthy adults. Exercise has been found to be particularly beneficial in terms of enhancing working memory for Met carriers. Greater levels of physical activity seem to offset the deleterious effect that the Met version of the gene has on cognitive horsepower.[21]

The increase in BDNF levels that exercise affords might benefit folks with dementias such as Alzheimer's disease, the most common cause of dementia in those sixty-five and older. Because Alzheimer's is characterized by a reduction in the number of neurons in brain areas such as the hippocampus, and because exercise helps support neurons in this part of the brain, exercise may help to slow the progression of the disease.[22] Indeed people with Alzheimer's who enroll in a twelve-week moderate exercise program show improvements in memory and brain functioning. After exercising, their brain worked more efficiently to do the same memory tasks they had completed before the program.[23]

Exercise might even be a successful preventative measure, taken like a vaccine to prevent the onset of Alzheimer's. Though intense aerobic exercise is particularly effective, you don't have to run miles on the track to keep your brain healthy; even washing dishes, cleaning, gardening, and cooking are linked to a reduced likelihood of developing Alzheimer's. Older adults who are physically active are often less likely to develop the dementia than their sedentary counterparts.[24]

The aerobic part of exercise seems key for improving mental health. The increase in blood flow that occurs when we swim, run, cycle, walk briskly, or even do household chores at a vigorous pace is an important part of propelling BDNF in the brain. Aerobic exercise is a catalyst for the appearance of the metabolic nutrients necessary to think sharply. Activities such as strength training and stretching don't result in the production of growth factors in the same way.[25] But when exercise is sufficiently aerobic, it actually alters the structure of our brain. The volume of the brain usually shrinks as we get older, and less brain means less power to think, reason, and do pretty much everything we need to do. But exercise slows this shrinkage. Recently neuroscientists found that the size of the hippocampus of older adults who walked for forty minutes around a track three times a week for one year increased by about 2 percent. Those who took part in a stretching routine instead showed the typical age-related decline in size of about 1.5 percent over the course of a year. Even later in life exercise can protect and improve the structure of our brain.[26]

Regardless of the specific cellular and molecular cascades created by increased exercise, there is a clear link between exercise and cognition. *Mens sana in corpore sano,* the ancient Romans wrote, which roughly translates as "A sound mind in a healthy body" and shows that the mind-body connection has been known for thousands of years. Exercise helps grow new connections in the brain and strengthen existing ones; older adults who exercise have the brains of much younger people; young kids who are the most physically fit score highest on important tests of achievement; and people who exercise regularly report worrying less and are less depressed than their sedentary counterparts. Though we are increasingly reliant on technology that keeps us in our seats, the power of movement is clear. The key to a better brain is rooted in the actions of the body.

Interestingly, it's not just exercise that alters the structure and functioning of the brain. Being sedentary can also change the makeup of the brain—though not necessarily for the better. In rats, physical inactivity has been linked to alterations in brain areas that are important for regulating the cardiovascular system, which sets up sedentary rats (and, likely, people) for hypertension and an increased risk of cardiovascular disease. Being active affects your brain in positive ways, and being idle affects your brain in unhealthy ways.[27]

Your activity not only affects brain functioning; it can also provide a window into how your mind is operating. Take walking. Doctors used to think that signs of slow walking were just a normal part of the aging process. They were wrong. It turns out that slow or unstable walking is often an indicator of subtle cognitive impairments. Many of the same brain circuits that control complex cognitive activities also help us coordinate the complex movements needed to walk down the hall. Using walking to assess cognition represents a real departure from how mental fitness is normally assessed in older adults: while they are sitting down. Many neuroscientists studying the aging mind firmly believe that, when older adults go to the doctor and get their eyes and their blood pressure checked, they should get their walking checked too. Even subtle signs that walking is slowed or impaired may tell doctors that something important is going on in the brain.[28]

Whether or not your goal later in life is to dominate on the track like Olga Kotelko, having a fit body is good for the mind, and it is also good for the pocketbook. Recent estimates suggest that reductions in hospital, nursing home, and home care costs associated with increasing physical activity as we age would save tens if not hundreds of billions of dollars each year.[29] Because keeping older adults physically healthy means that they are sharper mentally as well, encouraging structured exercise regimens may be one of the best ways to keep older adults independent longer. The end result would be less burden on their family, the health care system, and taxpayers. In the United States, there seems to be more emphasis on being mentally fit than being physically fit, so perhaps campaigns that emphasize the brainpower benefits of exercise will be more effective than those promising a perfect body. Of course, if an exercise regimen were a mandatory part of health care coverage—or at least provided big reductions in fees for those who participated—the savings could be as impressive as it would be to see Olga Kotelko run the 100-meter dash.

CHAPTER 10

Buddha, Alexander, and Perlman

USING OUR BODY TO CALM OUR MIND

Meditation from the Neck Down

Even though I've been doing yoga and meditating for the better part of a year, I'm still struggling to master a practice called Ujjayi breath, or cobra breathing. The goal is to breathe in and out through the nose deep into the throat, letting your diaphragm do all the work of controlling the speed and intensity of your breathing. When done right, Ujjayi breath makes a sort of rhythmic swooshing sound that conjures up memories of the ocean or the wind. My Ujjayi breath doesn't sound much like anything.

I've had countless teachers try to explain cobra breathing to me, and now I am hearing about it yet again. As I sit cross-legged in a simple wood-paneled room tucked into the lush jungle in Puerto Rico, I can't help but notice that my surroundings are certainly conducive to mastering Ujjayi breath and the calm mind that goes with it. Yet instead of finding a rhythm, I am starting to hyperventilate. My mind is all over the place, first wandering to the question of what I am going to eat for lunch after my meditation lesson and then to worries about all the work I left unfinished back in Chicago.

My friend and I arrived at the Casa Grande Mountain Retreat the previous night. Home to an old coffee plantation, the retreat has sleeping rooms and a meditation studio perched on stilts on the mountainside. Folks come from around the globe to take lessons from world-renowned meditation teachers like Jack, who is now coaching me through my breathing. In his late fifties and famous for promoting the idea that having a healthy body is a key to a healthy mind, Jack sees the body as a vehicle for changing the mind. He believes that having a fit body can actually enhance your ability to focus your attention, regulate your emotions, and even improve your memory. Meditative practices hone a connection between body and mind.

Of course, Jack is not alone in his focus on the body. Eastern cultures have long valued the body as much as the mind in promoting healthy well-being. Philosophers, artists, and playwrights also frequently tout the connection between the brain and the rest of the body. The poet Ovid said, "The mind ill at ease, the body suffers also." It works the other way around too: when you have trouble reining in your body, your mind tends to run wild.

Most of the time, we are thinking about what is *not* happening. We're reviewing past events or anticipating what is likely to happen in the future. Indeed this sort of "mind-wandering" is thought to be our brain's default operating mode. Although being able to think about what isn't going on around us can help us learn from the past and productively reason about the future, this lack of attention to the present comes at an emotional cost. Simply put, a wandering mind is an unhappy mind. People report being less happy when their mind is jumping all over the place rather than being calmly focused on the now.[1]

It turns out that there is something we can do to decrease our mind-wandering: meditation. Experienced meditators report less drifting in their thoughts during meditation practice than people who don't have a lot of meditation experience, and even when meditators are simply asked to not think of anything in particular, their brain does a better job of keeping them present-focused and in the moment. Meditation also teaches us how to control our body. Mindfulness meditation is a prime example of this body-centered focus.

Mindfulness plays a central role in many forms of meditation and usually includes two main components: attending to your immediate experience and having an attitude of acceptance toward this experience.[2] Critically these experiences are not centered solely in the mind; they also incorporate the body. Take a look at the instructions for two types of mindfulness meditation, concentration and choiceless awareness:

Concentration: "Please pay attention to the physical sensation of the breath wherever you feel it most strongly in the body. Follow the natural and spontaneous movement of the breath, not trying to change it in any way. Just pay attention to it. If you find that your attention has wandered to something else, gently but firmly bring it back to the physical sensation of the breath."

Choiceless Awareness: "Please pay attention to whatever comes into your awareness, whether it is a thought,

> emotion, or body sensation. Just follow it until something else comes into your awareness, not trying to hold on to it or change it in any way. When something else comes into your awareness, just pay attention to it until the next thing comes along."

As you can see, these techniques focus on the body as well as the mind. The idea is that, through a combination of body and mind training, you change both your physical and mental states. And it works, according to a number of studies. In one study, a group of neuroscientists from Yale, Columbia, and the University of Oregon asked several experienced meditators and a group of meditation novices to perform different types of mindfulness meditations while their brain was scanned using fMRI. Brain areas that are commonly active when our mind wanders are quieter in experienced meditators—while they're meditating and even when they are not. This is interesting because meditators' brains looked different when they are doing nothing at all. Meditation changes how the brain physically operates all the time.

At rest, meditators' brains showed stronger cross-talk between bits of cortex typically involved in mind-wandering and areas involved in self-control, specifically brain networks that help us keep what we want in mind and distracting information out, encompassing areas such as the anterior cingulate cortex and the prefrontal cortex. The brains of the meditators seemed to have developed the ability to automatically set off a warning signal when mind-wandering threatened to take over, allowing for the dampening of thoughts that could have led their attention astray. Meditation had transformed their brain so that even when they are not doing anything at all, their experience resembles a meditative state, a more present-centered state of mind. One key element of these practices is learning how to recognize what's going on, not just in your mind but in your body too, which trains you to rein in your wandering mind.

Of course, it's possible that the meditation experts the neuroscientists studied didn't learn to curb their wandering mind through meditation. Maybe, instead, these folks were attracted to meditation in the first place because they were endowed from birth with minds that don't wander as much as yours or mine. But a host of recent work showing the power of meditation—and especially the power of meditative practices involving the body—suggests otherwise. Meditation helps change our mind. For instance, mindfulness meditation can help alleviate anxiety and chronic pain and even reduce symptoms of obsessive-compulsive disorder. Mindfulness helps us to develop a heightened awareness of the present moment. By attending to your body and your thoughts nonjudgmentally, you can learn to think about your feelings as events that will soon pass, which limits the importance you attribute to them. When you don't get caught up in a cycle of worries, chronic anxiety and depression are lessened and you reduce the chances that you will experience emotional distress.

Mindfulness changes the brain in ways that lead us to distance ourselves from, well, our self. We all have a capacity for mindfulness to a certain extent. Training this capacity has a quieting effect on brain areas associated with our vigilance toward ourselves and the negative emotional reactions that often arise when we start focusing on events in the past or become wrapped up in the "what ifs" of the future. By considering thoughts and feelings as transitory mental events that can be separated from ourselves, we are less likely to worry, and positive health outcomes follow.

Given the benefits of meditation, it's no wonder that such techniques have become widespread, even outside of yoga centers and spas. They are now embraced by politicians, celebrities, and athletes, as well as health-conscious laypeople. When he was coaching Michael Jordan and the Chicago Bulls to several successive championships, Phil Jackson became well known for advocating meditation as a means of enhancing his players' performance. Successful individuals ranging from leaders of Fortune 500 companies to politicians have also touted the benefits of meditation practices in their everyday and work lives. The Dalai Lama himself has donated time and money to the study of the brain science of meditation; with the neuroscientist Richard Davidson at the University of Wisconsin, he is contributing to the founding of the Center for Investigating Healthy Minds. Davidson, who has practiced meditation for the better part of twenty years, has spent the past several years researching the power of mindfulness in adults and children, teaching meditation that focuses on both the mind and the body to fifth graders, for example.[3]

Changes That Last

A few years ago, researchers set out to document the brain changes associated with several months of juggling practice. That's right, the art of being able to coordinate multiple flying objects in the air so as to keep them aloft, not falling down at your feet, alters the brain physically. They found that when people devoted several hours a week to learning how to juggle, they showed changes in areas of the cortex involved in tracking motion. These brain areas became more dense with neurons, which generally signals better communication among brain cells. But when people stopped their intensive juggling practice, these motion-understanding brain areas that had grown more lush with practice thinned out again.[4]

In contrast, meditation's effects on the brain are long lasting. Researchers at the University of California, Davis, led by the neuroscientist Clifford Saron, recently put the longevity of meditation's benefits to the test and found that meditation works a lot like a vaccine.[5] You just need boosters every so often to reap the positive impact. Saron and his team invited experienced meditators to participate in a three-month residential meditation retreat at the Shambhala Mountain Center in the Colorado Rockies. This was not a part-time meditation study; people who took part in the program got at least five hours of training a day, so the researchers were careful to recruit people who had some meditation experience and knew what they were getting into.

More than a hundred people applied to take part in the retreat (and to complete the researchers' concentration tests that went along with it). Volunteers ranged in age from twenty-one to seventy and were from a variety of backgrounds. Thirty were selected to take part in the first retreat; another thirty participated in a second retreat right after the first. It was a clever idea to stage the study this way because it meant that Saron and his team could compare the minds of the first group to those in the second group, people who hadn't yet participated in the program. The second group, by the way, was flown to Colorado to take part in the concentration testing sessions with the first group in order to make sure that everyone was administered the tests under the same conditions.

B. Alan Wallace, a meditation teacher and Buddhist scholar, led the volunteers through their meditative practices. People learned to focus their attention on one stable aspect of their body, such as their breathing, and to recognize when their attention had wandered off and to rein it back in. At three points during the retreat—once at the beginning, once in the middle, and once toward the end—each person took part in a series of tests designed to measure concentration and vigilance. This amounted to watching a procession of lines move across the screen and signaling, by pressing a mouse, whenever they saw a line that was slightly shorter than the rest. This task demands intense and sustained concentration and is quite tedious.

As the UC Davis group mentions in the report of their findings, historical accounts from the Buddhist contemplative tradition describe meditation practices that are designed to improve sustained attention. So perhaps it isn't surprising that people got better at doing the attention tasks over their intense three-month meditation experience. This sort of improvement, importantly, wasn't seen in those volunteers who were simply waiting their turn to go to the retreat. As a group, the meditators exhibited superior attention skills.

Meditation also had an impact on the body. An analysis of samples of the meditators' blood showed a significant increase in telomerase, an enzyme that has been linked to being healthy and disease-free. Some cells in the human body constantly divide throughout life—skin cells and cells in our digestive tract, to name a few types. Others divide less often. The enzyme telomerase plays an important role in successful cell division, as it allows for the replacement of telomeres (short pieces of DNA located at the end of the chromosomes) during the dividing process. When cells divide, the end of the chromosome is often lost, along with the information it contains. Telomeres help protect these ends during division. As a result, healthy cells are maintained after division, not fusing with other cells or rearranging in such a way as to lead to abnormalities and cancer. Meditation may lead to emotional changes that help regulate this anti-aging enzyme.[6]

The findings about mediation's lasting impact are really striking. Five months after the retreat, laptops were sent to the participants' homes with instructions for administering the attention tasks to themselves. The scientists found that there was still a positive impact from the retreat. Those who had taken part in the retreat hadn't seemed to lose much ground from the last testing session in Colorado. The people who maintained the biggest boost were those who were still practicing meditation at home, even if they were doing it for only several minutes a day.[7] Booster shots of meditation seem to be enough to sustain the concentration benefits of intense meditation sessions.

Meditation training could benefit many professionals—air traffic controllers, pilots, even referees in professional sports—who need vigilance to successfully perform their jobs. A lapse of concentration, not minding the present, could mean missing a problem in a jet's path, missing the runway altogether, or missing an error or foul. Consider the pilots of Northwest Flight 188, who were in a heated discussion about airline policies and overflew their destination by more than a hundred miles.[8] Perhaps meditation training would have ensured that the pilots stayed on task.

Back at the Casa Grande Mountain Retreat in Puerto Rico, as I was finishing my meditation lesson, Jack said, for what may have been the tenth time in our two hours together, "Focus on your body posture, on how you are sitting, on your shoulders and back. Your mind will follow." Jack teaches a combination of meditative traditions, but his focus on the body reminds me of a relatively new meditative practice (at least new to the West) called integrative body-mind training (IBMT) that scientists have recently been showing has a powerful impact on the mind.

Adopted from traditional Chinese medicine, IBMT incorporates body relaxation, mental imagery, and mindfulness training, guided by a coach and assistive CD. It puts a special emphasis on cooperation between the body and mind and stresses a state of "restful alertness."[9] The idea is that a successful meditative state is achieved through an optimization of connections between mind and body, particularly, that the body has the power to change the mind. In IBMT, people learn to form a high degree of awareness between body and mind. IBMT does not stress efforts to control thoughts; you adjust your focus gradually through awareness of posture and body relaxation so that unwanted thoughts are less likely to co-opt your attention and distract you.

You don't have to be an experienced meditator to enjoy the benefits of IBMT. A recent study showed that it can change our brain for the better with eleven hours of practice, even if we have no previous meditation experience. Neuroscientists observed improvements in the nerve fiber tracts connecting frontal areas of the brain such as the anterior cingulate cortex to other brain structures after eleven hours of IBMT conducted over a one-month period. The ACC is part of a brain network important in the development of self-control and emotion regulation.[10] In another study, just five hours of IBMT over a two-week period led to a 60 percent reduction in smoking among a group of smokers.[11]

As someone who is always pressed for time, I was definitely interested in these sorts of results. But I did wonder how such short bouts of IBMT could be so effective, so I did some research and came to the conclusion that IBMT seems to be a lot like riding a bike. When you are a child and first get on a bike, you tend to pay attention to every aspect of your performance—how you are balancing, holding the handlebars, what you are doing with your feet, arms, and hands. This constant vigilance requires a heavy dose of input from the prefrontal cortex (where our focus of attention is largely housed). But as you get better and better, you reach a point at which all of a sudden you don't have to pay attention to everything you're doing. Indeed studies of activities like bike riding or hitting a golf ball reveal that, as people improve, these activities are performed largely outside conscious awareness. Something similar is thought to happen in meditation practices such as IBMT. There are noticeable changes in brain states that go hand in hand with meditative experience, one being less involvement from brain regions involved in consciously directing our attention.[12] Short bouts of IBMT might do the trick because people begin to automatically deal with a wandering mind, even when they aren't meditating.

IBMT is all about progressing from a highly self-conscious practice to an effortless one. Strengthening the connections between the prefrontal cortex and the rest of the brain allows our mind to run on autopilot.[13] While they are in a deep meditative state, practitioners totally forget the self and the body. The prefrontal cortex takes a break.

Whether you want to hit a five-foot putt to win the tournament, nail a stress-filled pitch to a client, or ace an important test, meditation may help regulate your thoughts, emotions, and behaviors when you need to perform at your best under stress. Al Gore and Hillary Clinton, two politicians who are constantly performing under pressure, have attested to the power of meditation in helping to reprogram their mind. They use meditation practices as a way to rein in their wandering mind and combat the stress of living in the public eye. And it doesn't take years of practice. Even short practices with techniques such as IBMT can improve your thinking and alleviate stress.

When people are under stress, cortisol (a hormone produced by the adrenal gland) increases and is associated with stress-related changes in the body, such as higher blood pressure and a quick burst of energy. Measuring the concentration of cortisol in blood or saliva can provide reliable information about how stressed a person is at a particular moment. When psychologists want to study how reactive to stress people are, they use a test in which they ask subjects to do some mental math out loud (say, continually subtract 47, beginning at 1,934, as quickly and accurately as possible). In most people, this test produces a sharp increase in cortisol concentrations in their blood or saliva. This spike in cortisol can be reduced by just a few weeks of integrative mind-body training.[14]

IBMT also induces changes in self-control when people aren't trying to focus on anything in particular. Indeed when researchers looked at the brains of smokers when they were simply resting after their meditation training, they found increased activity in brain areas such as the anterior cingulate cortex, brain areas related to self-control. Meditation training may make self-control easier and more automatic—a real coup for smokers who are trying to quit and have to rein in their impulses to act on those urges to light up.

There seems to be real power in meditation like IBMT that combines both physical and mental training. If half a dozen to a dozen hours of meditation can change the brain and help enhance performance on the playing field, in the classroom, or in the boardroom, then it may be time to rethink our weekend activities. After all, that's less time than it would take to watch four football games or repaint the bedroom.

Musical Bodies and Minds

Every Tuesday for the past six months I have taken a violin lesson in the River North area of Chicago. It's actually a pretty funny scene. I go in for my 3:30 lesson just as a five-year-old kid and her mother come out. When I leave my lesson, another five-year-old enters, parent in tow. It seems as if the mothers and fathers of these musical kids are always wondering where my child is; it's not until they see me holding the violin case that they realize I am the one taking the lesson. I played the violin from age eight until I was eighteen. I stopped when I went off to college and got immersed in other life pursuits. Now, as a thirty-something adult, I have come back to music, struggling to learn finger positions, bowing, and stance, activities that as a kid I never thought much about; they came so fluently to me then.

The first few months playing the violin again were the most frustrating. I came to each lesson eager to make progress in terms of the pieces I was playing; I wanted to go from simple scales to Bach minuets. But I made little progress. It wasn't rare to spend an entire lesson on my stance or the placement of my fingers around the bow. I was growing impatient. It was then that my teacher, Jenny, explained that she was coaching my body first, before she taught me the music. This body-centered focus makes a lot of sense when considered in the context of meditation techniques like IBMT. By training my body, I would develop the control and balance needed to play concertos. We rarely think about how great performances occur, how our body gets us there. Jenny's idea was that putting my body in the right position would actually make it easier for me to understand the music.

Teaching the body also frees up the prefrontal cortex and the working memory housed there to play sonatas and concertos at the highest possible level. Just as in learning to ride a bike, first you must devote a lot of working memory and conscious control to how you are holding your body. With time, however, these movements become more automatic and habitual in such a way that your brainpower is available to devote to musical theory and interpretation.

Molding the body as a first step in teaching the violin may be especially effective for young kids, given that cognitive horsepower develops with age. Because children have less working memory than adults, getting the motor parts of playing down pat allows them to use all their conscious control for musical interpretation. And because the prefrontal cortex isn't thought to reach full maturity until well into early adulthood, other brain areas like the sensory and motor cortex have a lot of input in what is learned early in life.[15] Thus teaching the motor aspects of music first, especially when we are young, may be beneficial, as the areas of the brain that primarily control movement are poised to take in information and help us commit it to memory.

Most great violinists play beautiful music that resonates from their instrument, but their body is often involved too. Even masters such as Itzhak Perlman, who contracted polio at age four and plays sitting down, moves his body in fluid and balanced ways during his performances. The role the body plays in musical expression is one reason why teaching that helps people learn to control their body is so popular in music schools. Take the Alexander technique, which music teachers think of as "an owner's or operation manual which helps students to re-educate and restore beneficial postures and movements."[16] Founder of the mind-body method that took his name, Frederick Matthias Alexander was born in Tasmania in 1869. In his thirties he emigrated to England, where he lived and worked as an actor for most of his life. His acting career was almost cut short when he began developing unexplained symptoms of laryngitis, losing his voice with the stress of an upcoming performance. He visited several doctors to find a cure for his problem, but they were of little help. It wasn't until Alexander examined his body posture in a mirror one day that he noticed something unusual: whenever he was about to speak his lines, he would tighten his neck muscles, pull his head back, and suck in air through his mouth. This sort of posture couldn't have been good for delivering his lines, and he had a hunch it might be responsible for his voice loss too. Through careful observation of his own body movements, Alexander eventually taught himself how to loosen the tightness in his neck and head. Lo and behold, his attacks of laryngitis also went away. Amazed at the power of reeducating his body, Alexander started teaching his newly discovered kinesthetic sense to others and called it the Alexander technique.

Alexander noted, "You translate everything, whether physical or mental or spiritual, into muscular tension."[17] Retraining people to incorporate breathing and a body-centered focus into their daily lives could change how they think.

The Alexander technique is a mandatory part of training in many leading music, theater, and dance schools around the world, which believe that it not only improves musical and technical skills but also lowers stress and anxiety in performers.[18] Learning how to control the body, to quiet tension and stress, can have a profound impact on the mind. Michael Langham, director of the Juilliard School in New York, has commented, "Alexander students rid themselves of bad postural habits and are helped to reach, with their bodies and minds, an enviable degree of freedom of expression."[19]

The power of the Alexander technique extends beyond the

musical world. We often think that our poor posture or back problems or how we carry ourselves when we walk is a part of who we are, something we inherited. But we actually develop many bad movement habits through the types of repeated activities we do every day. Sitting at the computer, we hunch over the keyboard, shoulders at our ears, motionless. Though people once moved around the office constantly, to make a copy, send a fax, or get a drink of water, most of these activities have now been replaced by the click of a button, an email or a text, or a large water bottle. This sedentary lifestyle can have a negative impact on our body, but it also takes a toll on our mind.

Many workplaces recognize that you can reduce strains—tight neck and shoulders, sore wrists, and lower back pain—that accompany sitting for long hours with a good chair and the appropriate desk. But no matter how high-tech your workplace is, if you sit slumped at your desk all day, your body will show signs of tension and stress. The Alexander technique teaches that the only way around these strains is to have a better idea of what your body is doing. Don't ignore those aches and pains in order to get the job done, but consider them to be alarm signals that you need to take action. Just as Alexander changed his posture and regained his voice, you can develop body awareness that helps you feel better physically and mentally. Through touch and gentle body guidance, Alexander teachers help you become aware of how you perform everyday actions and take the tension out of them. Sit upright, tap the keyboard lightly, and release neck tension—these simple adjustments can go a long way to help you be more in control of your entire body and mind. Teachers of the Alexander technique argue that simple adjustments like these will not only lead to more comfort physically but that you will be better able to operate mentally as well.[20]

I think of the Alexander technique as an extreme form of integrative mind-body training. You gain a high degree of awareness of your body, which helps you focus on what's important and feel better; it can even help lift you out of a depressed mood. Knowing your body is key to getting control of your mind and your performance.

The Alexander technique can help reduce back pain.[21] It has also been shown to help people with Parkinson's combat both their movement disorders and mental disorders such as depression, which affects approximately half of those who develop Parkinson's disease. Depression is not just a reaction to having this disease. Parkinson's is often related to changes in the production of neurotransmitters such as dopamine. When dopamine levels circulating in the brain are drastically reduced, movements become shaky and feelings of anxiety and depression can occur.[22]

A few years ago, a group of researchers in London asked patients diagnosed with Parkinson's disease to take part in a three-month study of the Alexander technique.[23] A teacher used hands-on techniques to help patients learn how they can control their movements and balance in everyday activities. There were two control groups, one of which didn't get any treatment, and another that got massages instead of Alexander lessons. Patients were randomly assigned to one of these three groups, a crucial part of any valid study, but a protocol that wasn't always followed in previous studies on the Alexander technique.

Both before and after the study, the London researchers took a battery of measures designed to establish the patients' basic motor skills and mood. People were asked to rate how easy or difficult it was to perform actions such as walking, getting dressed and undressed, and turning over in bed—both when they felt at their best and also when they felt at their worst. They also filled out several common measures of depression. At the end of the study, patients who had taken part in the Alexander technique training reported that it was easier to perform daily activities. Most striking was that people who got the training were also comparatively less depressed than patients in the control groups.

Other researchers have noticed a link between movement disorders such as dystonia, characterized by involuntary jerks and tremors, and depression. An inability to control the body propagates up to the mind, making it hard for folks to control their negative thoughts and feelings. But learning strategies to take back their body (or at least be more aware of what they're doing with their body) helps to change this.

Given that changes in the body, such as moving from a slumped posture to sitting upright, can positively affect how you feel, it seems logical that learning how to better control your body, how to move and act in a way that lessens pain and encourages fluidity and balance, can also change your ability to think. Our mind has a tendency to be restless and wander. Many think that this tendency extends to the body. In other words, the way we operate in the world puts our body in a tense and restless state, yet we have the power to get our body back—through exercise, meditation, IBMT, and even the Alexander technique.[24]

When you think of your body as a shell, merely the casing for your brain and less important than your mind, you will be less healthy. With this perspective, you can't really take proper care of your body. In fact, the health of your mind is profoundly linked to that of your body. Knowing this, you'll make healthier choices in your life about what to eat, when to sleep, and how to behave.

When you appreciate the power of the body in changing the mind, you function better.[25] Exercise and body-centered meditation, awareness, and learning practices that coach the body as well as the mind can help you achieve this mind-body connection. Our thinking extends way beyond the cortex.

CHAPTER 11

Greening the Brain

HOW THE PHYSICAL ENVIRONMENT SHAPES THINKING

Beyond the Body

Blurry-eyed, I stumble out of bed and pull open the window in our tiny bedroom. Gazing out over the lush garden I realize that I have no idea what time it is. Even more striking, I find myself not really caring. There is no clock in the room, and my cell phone, which doesn't work here anyway, has died. The only timing aid at my disposal is the sun hanging high in the sky.

It's a small thing, not caring what time it is, something vacationers often do when they relax, unwind, and lose track of the outside world. It usually happens to me around the third day of our annual pilgrimage to the small Italian seaside village of San Felice Circeo, where my husband's parents have had a summer home for the better part of forty years. Dario grew up in Rome and, as a boy, spent summers in San Felice, where, according to him, nothing has really changed. This is especially true when it comes to technology, he moans.

There is no computer in the family's Italian villa, no wireless network to plug your laptop into, and email is accessible only at the town's one Internet café, a twenty-minute drive away. Of course, even when you make it into town there is no guarantee you will actually be able to get online because half the time the network is down and the other half the elderly *nonna* who owns the café has, with no forewarning, closed shop to meet with friends. In short, vacationing in San Felice is like a trip into a wireless abyss, something that is increasingly rare today, when people can get "plugged in" in even the most remote areas. Yet after a few days of being separated from my email and existing without regular reminders of the time, I find myself less concerned about what I am missing than if I had perpetual access to the outside world.

Perhaps losing that nagging feeling to constantly check your phone is not something most people would think twice about, but as a brain scientist who makes her living studying the inner workings of the mind, I am particularly interested in what happens when we step away from the distractions of our daily life. I want to know how our surroundings—from the technology that keeps us permanently online to the general chaos of urban life—can affect our ability to focus our attention, make decisions, or even learn something new.

In recent years, neuroscientists have realized that our environment influences our mind in unexpected ways. In fact our mind extends beyond the physical cortex of the brain's flesh. Our brain is not the only resource we have at our disposal to reason or solve problems. In other words, we have to change what we think of when we think of *cognition*.

Scientists no longer view the body and its surroundings as

simply the foil or casing for the mind. Nor do we view cognition as the sole driving force behind what the body does. Striking new evidence makes it clear that particular types of bodies (fit versus more sedentary) and the specific way these bodies act have enormous consequences for how sharp our thinking is. Moreover, it's not simply the body per se that has a significant impact on the mind. The environment that this body is in contributes nontrivially to our thinking and reasoning skills.

For instance, starting off the morning by having to navigate through intense city traffic or by receiving a dozen emails with "urgent" in the subject line can decrease your ability to shine in a meeting later in the day. Whenever your attention is dramatically captured by stressful situations, your thinking changes. Neurons go into crisis mode. Neural areas involved in focusing attention buckle down and stop communicating effectively with the rest of the brain. This makes it hard for different areas of the brain—those involved in logic, memory, or attention—to work together to help us function at our best.[1] Perhaps most interesting, this crisis mode doesn't stop when the stressful situation is over. Just as it takes time for our body to recover after physical exertion, it takes time for our mind to recover after mental exertion. It might seem quite alarming that operating in a hectic environment one minute can adversely affect the brainpower we bring to the next. But it works the other way too: spending a few hours in the garden can increase the likelihood that you'll perform at your best long after you have gone inside to a home office or business appointment.

Below are a couple of thought tasks for you. Take out a blank piece of paper and a pen and do the following:

1. List all the uses you can think of for a brick.

Next,

2. What would be the consequences if everyone suddenly lost the ability to read and write? List as many as you can foresee.

What predicts how many different solutions you will be able to come up with? Your motivation or knowledge of the world might come into play, but you might be surprised to learn that another factor is what you were doing before you encountered this exercise. What you do in the middle of problem-solving matters too. If you are able to take a break, you are more likely to come up with new answers to problems like the ones above.[2] Walking away from a puzzle or challenge brings new possibilities to the surface and also flushes out dead-end thinking. It's akin to rebooting your computer when it freezes. Stepping back helps get rid of bugs, creating new opportunities for insight.

I often describe this phenomenon of walking away in talks I give to companies about how we can use findings from brain science to improve our daily lives. People find it counterintuitive that stepping away from what you are working on increases the probability of success. My guess is that this is because we have a tendency to want to avoid doing anything that seems to initially take us farther away from a goal—backup avoidance, psychologists call it. But if you think about it for a moment, it's easy to come up with instances when turning your back on a problem helped you solve it.

After a recent presentation I gave, Gary, a computer programmer who works in downtown Chicago, came up to me and shared a story. As Gary explains it, whenever he hits a roadblock at work, no matter how long he bangs his head against the wall trying to solve the problem, the answer always comes to him on his walk home to his house in the Oak Park suburbs at the end of the day. The bus drops Gary off right by Oak Park's Austin Gardens, a picturesque park and nature preserve. Walking through Austin Gardens, Gary told me, he has solved more programming issues than anywhere else.

Neuroscientists have known for some time that, when rats try to solve a problem like navigating through new surroundings, their neurons fire in new ways. And it's when the rats take a break from their exploration that they convert these new firing patterns into long-lasting memories of the experience.[3] Something similar seems to apply to how humans reason and learn. It's only when Gary takes a break that the solution to his coding problem pops into his head. Interestingly, it's probably not just the act of stepping away or a change in environment that improves his thinking. Being more likely to come up with answers when he is in the park resonates with some remarkable discoveries scientists have made about how nature affects our brainpower.

Nature, it turns out, can have a powerful influence on our thinking. Poets, writers, and philosophers have long suspected that spending time outdoors offers positive benefits for our health and well-being. Being in harmony with nature is a central idea in many Eastern cultures and practices. The practice of Tai Chi, martial arts, meditation, and yoga often occur in parks. In the West, national parks attract millions of visitors every year, and many people who want to relax and unwind choose to hike in the mountains or walk on the beach. The belief in nature's restorative properties can be seen in how we pick our vacation destinations and choose our extracurricular activities.[4]

Brain scientists have recently discovered that the positive benefits of nature extend beyond our physical health to our mental capacities. Frances Kuo, who runs the Landscape and Human Health Laboratory at the University of Illinois, studies the relationship between people and the physical environment. Her research reveals that the impact of nature on the human psyche goes beyond nature's aesthetic appeal. Kuo has found links between green space and a safe home life. She has also discovered that natural surroundings are tied to enhanced working memory, which translates into increased concentration and self-control.

Kuo is particularly interested in how nature affects well-being in inner-city settings. There are many demands on your attention when you live in poverty. "Basic concerns such as rent, utilities, and food are ongoing challenges that require effortful problem solving and reasoning," Kuo writes.[5] Safety concerns, both within the family and outside it, mandate a level of vigilance that the middle class and well-off can't even comprehend. Surviving in the inner city requires a high degree of self-discipline that, at the most basic level, comes down to what psychologists call executive control.

Executive control is an umbrella term that refers to a collection of cognitive functions, such as the ability to focus attention and working memory, which helps us keep thoughts we want in our consciousness and unwanted thoughts out. When we can't successfully manage our impulses—whether it's reacting violently in an argument, gambling away the rent money, or giving in to eating that donut we swore we were going to avoid—a failure of executive control is usually lurking in the background.

In one study, Kuo asked residents of the Robert Taylor Homes, a housing project in inner-city Chicago, to think back to situations in which they had a disagreement with a family member and how they dealt with that conflict. Were they able to reason through their differences and talk it out? Or did the altercation end in physical violence or even the use of a gun or other weapon?

A typical resident of the Robert Taylor Homes whom Kuo studied is a single, thirty-four-year-old African American woman with a high school diploma or equivalency degree. She's raising three children on a household income of less than $10,000 a year, which she might make in a variety of ways, such as part-time work at a local fast-food chain or nearby convenience store. Not only is making ends meet a constant struggle, but keeping peace in such a stressful environment is very difficult, mandating a high degree of executive control.

Some of the residents of the Robert Taylor Homes live in relatively green high-rise apartment buildings, meaning that they have views of trees and grass out their windows, while others live in more barren surroundings, with views of, say, a vacant lot. Kuo had a hunch that what people can see outside their windows might have a real impact on their ability to manage the stress in their home. She found that the more green residents could see outside their windows, the less aggression and violence they reported at home.

In the early 1960s, when the Robert Taylor Homes were first built, bushes, trees, and grass were planted around each of the twenty-eight high-rise buildings in the complex to give them some of the characteristics of a typical suburban neighborhood. With time, however, many of the green spaces have been destroyed and paved over in an effort to keep down maintenance costs. Yet the leftover patches of green still have a positive effect on what happens in residents' homes.

Can it really be the case that seeing green space outside one's window leads to less violence? Or could it be the other way around, that less violent households are rewarded by being assigned to greener buildings in the first place? Kuo is convinced that the former is true, because, when people apply to the Chicago Housing Authority for placement in one of the seventeen projects located throughout the city, they have no say over where they live within a particular development. As Kuo points out, clerks in the central office handle all the housing assignments, which involve some forty thousand residents spread out across 1,500 buildings in the city. No bureaucrat can remember the characteristics of so many buildings, let alone take them into account when assigning apartments.

Kuo also measured the residents' level of executive control, specifically the amount of working memory they had. Working memory is largely housed in the prefrontal cortex, the very front part of the brain, above the eyes. The prefrontal cortex is what really distinguishes us humans from other animals. Not only is the human prefrontal cortex much larger than that of other primates of similar body size, but it occupies a far larger percentage of our brain than it does in any other animal. The prefrontal cortex is the seat of many of the mental abilities that make us uniquely human, including self-control, and it also plays a big role in regulating emotions. The more powerful your prefrontal cortex and the working memory housed there, the better you are at not letting your emotions get the better of you.

Kuo found that residents who had views of nature scored higher on tests of working memory than those who had barren views. And the more working memory, the less violence they experienced in the home. Even a small area of a natural environment outside your window changes working memory for the better and gives you the self-discipline to keep your own emotions in check and effectively handle altercations that occur in your home.

The benefits of green space are widespread. University students with mostly natural views from their dormitory room score higher on tests of working memory and concentration than college students who live in the same dorm but with views of other buildings.[6] Similar to the Robert Taylor residents, university students rarely get to choose exactly where they live. They may get to specify their preference for a particular area of campus or a set of dorm buildings, but they usually don't get to choose the particular room. So it's not that the students with better concentration skills opt for a greener view, but that the views affect students' concentration abilities.

Kuo's work shows that something as simple as being in nature, even looking out at a patch of it from indoors, can help boost working memory, focus attention, and get things done. This is good news for parents of children with attention-deficit hyperactivity disorder, because central among the deficits that characterize ADHD is an impairment in working memory, which contributes to kids' inability to control their impulses and behavior.[7] If exposure to green space helps boost working memory, then being in nature should help curtail some symptoms of ADHD. And it seems to. Studies have found that parents rate their children with ADHD as functioning better than usual after having engaged in activities in green settings relative to indoor or even outdoor city settings.[8]

That nature improves brain functioning was known by even the first psychologists. In the late 1800s, William James made a distinction between two types of attention. Certain elements in the environment are effortlessly engaging and draw on *involuntary attention*: "strange things, moving things, wild animals, bright things," James wrote.[9] In situations that don't effortlessly engage us, we need to execute voluntary or *directed attention* instead.

Scientists compare directed attention—which is at the heart

of our ability to concentrate—to a mental muscle that can wear out over time. When we're in nature, our surroundings (whether the sound of a bird chirping or the sight of a beautiful sunrise) attract our involuntary attention, which gives our directed attention, which is fueled by working memory, time to rest and replenish. If we never give this focus a break, it deteriorates.

Being in the middle of a crazy urban environment has the opposite effect. Cities are filled with objects and events that capture involuntary attention from one moment to the next: the horn of the car that is just about to run you over, the ringing bell of a bicycle messenger, the blaring alarm of sidewalk cover openings, and the rattle of the granny cart and baby buggies in your path. You also have to use your directed attention to consciously steer clear of advertisements that aim to lure you into buying something you don't want or need. In short, urban environments are a lot less restorative than natural ones.

Researchers at the School of the Built Environment in Edinburgh, Scotland, asked volunteers to take a walk that wound through both built environments and natural landscapes while they wore a mobile EEG device that captured their brain waves. The researchers found reductions in patterns of brain activity associated with being aroused and engaged (that is, reductions in directed attention) when folks walked out of the city and into green space.[10] Akin to James's observations about nature, when people are walking through the park, their brain is, in a way, quieter than when they travel in urbanized and busy areas.

Professor Stephen Kaplan at the University of Michigan has given a name to James's observations about nature: attention restoration theory.[11] In a series of cleverly designed studies with his colleagues Marc Berman and John Jonides, he has put James's ideas to the test. In one study, students were asked to perform tests designed to measure their directed attention abilities. First, they had to listen to letters of the alphabet presented in random order, memorize them, and then recall them in the opposite order. An experimenter sat next to them and wrote down their answer. This type of task is quite difficult because you have to keep moving items in and out of your focus of attention.

Second, students were asked to take a fifty-minute walk through either the Ann Arbor Arboretum or through downtown Ann Arbor. They did not get to choose which walk they went on. Both walks were 2.8 miles long, mapped out ahead of time, and everyone wore a GPS watch to ensure that they followed the prescribed route. The Arboretum is largely tree-lined and secluded from traffic and people. In contrast, those who took the downtown walk were led down traffic-heavy Huron Street, flanked with university and office buildings. After all the students got back to the lab, they performed the computer tasks again.

Third, a week later, the students returned and repeated the whole procedure, except that they walked in the environment they had not initially walked through. The results were clear. When students walked through the Ann Arbor Arboretum, they scored better on the tests of directed attention after their walk than before. The scores of students after walking through downtown Ann Arbor did not improve. Some environments simply bring out the best in people.

In fact, you don't actually have to take a walk in the woods to reap nature's brainpower benefits. In another study, Kaplan and his colleagues found that simply spending ten minutes (yes, just ten minutes) looking at pictures of Nova Scotia scenery improved people's concentration compared to when they looked at pictures of cityscapes from Ann Arbor, Detroit, and Chicago. Just as Kuo found that being able to see greenery out your window increases cognitive functioning, gazing at pictures of nature offers many of the same benefits.

The ability to concentrate is important because it enables people to mentally buckle down and stay on a task long enough to make progress. By mildly engaging our involuntary attention and letting those directed attention skills have a rest, we allow nature to rejuvenate the very cognitive processes that are so important for performing at our best. This new research tells us that nature's effects aren't simply peacefulness or quiescence alone. Rather, nature's ability to modestly capture our involuntary attention and give the rest of our brain a break creates a big impact on how we function.

Interacting with nature may have the biggest impact for people who are down in the dumps. Walking in nature leads to larger gains in working memory for people with symptoms of depression compared to healthy folks. Mood also improves after a nature walk. Interacting with nature has even been shown to have cognitive benefits for women with breast cancer who are burdened with worries about their cancer, medical treatments, and longevity. The mental fatigue that goes with depression and major illness can be combated by being in a natural, restorative environment.[12]

City Living

Humans evolved in natural environments, and we seem to thrive in them. Yet it's estimated that by the year 2050, 69 percent of people in the world will live in urban areas. There's no denying that there are many positive aspects of city living, such as easy access to health care, food, and other services (at least for some), but there are some downsides too. City living tends to be socially stressful; whether it's competing to get into the best schools and top restaurants or trying to find adequate housing or simply an open parking space, people who live in urban environments constantly battle with others for limited resources.

Scientists speculate that this battle of city living alters the human mind—and not necessarily for the better. Meta-analyses, in which scientists have aggregated the results of hundreds of studies, show that city dwellers are at a 20 percent increased risk for developing anxiety disorders and a 40 percent increased risk for mood disorders compared with people who live in less populated areas. Even more striking, the incidence of schizophrenia is almost doubled in people born and brought up in cities relative to those who were not. Simply put, on average, people have more mental distress and lower well-being when they live in urban areas with little green space. Of course, a link between urbanization and mental health doesn't necessarily mean that city living in itself is the problem. There could be some other factors that push people toward an urban lifestyle and also contribute to mental health problems. But scientists like Andreas Meyer-Lindenberg at the University of Heidelberg in Germany think there is good evidence to show that city living itself does the damage. Using schizophrenia as an example, Meyer-Lindenberg points out that there is a large dose-response relationship for the occurrence of this brain disorder, meaning that the longer you live in a city, the more likely you are to develop schizophrenia. It's hard to explain how the length of time you've lived in a city could be systematically related to your chances of developing schizophrenia without implicating the city itself as the driving factor.

A few years ago, Meyer-Lindenberg and his research team set out to better understand the link between urbanization and the brain.[13] The scientists began by inviting volunteers from a variety of backgrounds—some who were born and raised in big cities, others who lived in small towns—to have their brain scanned. During scanning, the volunteers performed a horribly difficult set of math problems, designed to ratchet up the type of stress that often accompanies city living; average scores were between 25 and 40 percent. The volunteers wore headphones in the scanner so they could hear the researchers tell them they were failing miserably.

Having someone repeatedly point out that you are screwing up is a surefire way to induce social anxiety in most people, and those in the study showed big spikes in their cortisol level after finishing the math problems. More surprising, Meyer-Lindenberg and his colleagues found that people who currently lived in cities showed increased activation in the amygdala (relative to folks who lived in small towns or rural areas) while they were being berated by the researchers. As noted in previous chapters, the amygdala is a major player in our emotions; among its many functions, it signals environmental threat. It has also been implicated in anxiety disorders, depression, and violence—all of which are increased in cities relative to rural areas. Increased amygdala activity often coincides with unpleasant emotional reactions. Simply put, living in an urban environment goes hand in hand with an increased sensitivity to social stress.

The researchers found that where participants had grown up also mattered in their reactions: the more densely populated the area, the more active their anterior cingulate cortex was during the stressful performance situation. Like the amygdala, the ACC is highly involved in emotional reactions; indeed it works closely with the amygdala to help make sense of our emotions. But the ACC has another function too: when something goes wrong, it emits a neural alarm signal that lets the rest of the brain know that something is amiss. People who had grown up in the city and currently resided there showed the most amygdala and ACC activation in response to induced stress.

To make sure the results were not a fluke, Meyer-Lindenberg and his colleagues ran the study again with a new group of volunteers. They even ratcheted up the intensity of social stress: the entire time the volunteers were doing the math, they could see a video feed of a disapproving investigator watching them fumble. This time the volunteers didn't just have to listen to someone comment on their failure; they had to watch him as well, grimacing and frowning every time an answer was wrong. In this second group, too, city living was associated with increased activity in the amygdala and ACC under stress.

A skeptic might argue that these findings have nothing to do with social stress and everything to do with the fact that people had to perform a demanding cognitive activity—solving math problems—in the scanner. Maybe city folks just show more activation in emotional brain networks when they perform difficult tasks. But the researchers took care of this criticism by scanning volunteers' brains while they performed the math task alone, without the disapproving experimenter. When there was no social stress, the link between city living and brain activation disappeared.

Interestingly, the amygdala and ACC aren't associated just with stress; they also relate to our social network size. The bigger these brain regions, the larger and more complex our social networks are. Because the amygdala and ACC are major players in our emotional reactions, it makes sense that they would be at the center of a brain network important for socializing, helping us recognize whether somebody is a stranger or an acquaintance, friend or enemy.[14]

Perhaps city living and the varied social interactions that come with it furnish city dwellers with bigger brain equipment to deal with the complex situations they face. If so, this would be consistent with something known as the "social brain hypothesis," the idea that through evolution, living in larger, more complex social groups led to a selection bias for larger brain regions with a greater capacity for performing important social computations, such as learning who is who and remembering many faces and relationships. Across different primate species, those who live in larger social groups tend to have a bigger amygdala relative to those who don't, even when controlling for overall body and brain size. Of course, people better equipped with the brainpower for socialization might travel to urban environments where social interactions are prevalent. Regardless, with bigger equipment also likely comes an increased risk of malfunctioning. Because city dwellers constantly use this equipment to deal with difficult social situations they frequently encounter, the equipment may stop working the way it should, becoming hypersensitive when even mild forms of stress come its way.

Our interactions with others not only affect our responses to stressful situations; they can also affect how we feel about our ability to face challenges. As we saw a few chapters back, when we are out of shape, our lack of energy impacts how difficult we think various activities will be to perform. When people are asked to estimate how steep a hill is, they view the incline as more extreme when they are out of shape or wearing a heavy backpack. We get information from our body about how demanding it will be to walk up the hill, and with this information we judge its physical qualities. The same thing happens when we are psychologically depleted—when we have had a stressful interaction, a fight, a heated argument, or just think about someone who has betrayed or disappointed us. Having a friend close by—or just thinking about a supportive friend—can change how difficult we perceive it will be to face a challenge; people judge hills as less steep and difficult to traverse when they are accompanied by a friend. However, this power of friendship depends on the quality of the relationship. The longer you have known the person, the closer you are, and the greater the interpersonal warmth, the more being next to her or just thinking about her lessens the impact of the hill. When you think about someone toward whom you feel ambivalent, you will see the hill as steep and treacherous to climb.[15]

Perception of the physical world is not determined solely by the environment itself, such as how steep a hill actually is, but is also shaped by how much energy it would take to negotiate the space or situation. When our physical resources are depleted (due to age, fatigue), hills appear steeper; when we are psychologically taxed, the same thing occurs. When we have friends around, however, the situation changes. Social support changes our mind and eases our judgment about a task's difficulty. It also reduces our physical reactivity to stress. For instance, the cardiac stress reaction that is often created by challenging mental tests is smaller when you are accompanied by a supportive person than when you are alone.

This notion that social support can change how we think about physical challenges—that our psychological feelings of inclusion and happiness can seep into and affect how we feel about facing physical challenges—supports the view of depression as an inability to change the physical world. Indeed depression often goes hand in hand with learned helplessness, a phenomenon in which people (and other animals) don't feel they have control over a particular situation or outcome, so they stop trying to reach a goal. When you feel psychologically helpless, physical challenges, such as getting out of bed, loom larger than they are. Perhaps helping depressed individuals act, to physically move, can help alleviate the depression.

Changing how the body moves in the world is one way to do this. As we saw in the first chapter, people who have had Botox treatments, who can't move their face as easily into a frown, show fewer depressive symptoms and are slower at understanding negative information than folks who don't have this facial paralysis. If depressed individuals can be encouraged to move around and to act in ways that contradict the feeling of a lack of control in their psychological life, they may feel better mentally. When people feel as if they can't take a step forward to put their life back on track, actually taking steps—putting one foot in front of the other—may be advantageous. Our brain doesn't always make a distinction between a physical act and a psychological emotion, so experiencing physical control should lead to increased feelings of psychological control.

Feeling depressed or sad is often described as feeling *down* or *low*. Recent research suggests that depressed individuals are more likely to orient their attention downward in visual space compared to people who are not experiencing this negative emotional state. The very act of looking up or standing upright might very well help to lessen depressive symptoms.[16]

Encouraging folks who feel down in the dumps to think about how they can physically act differently might help too. A study conducted with Canadian Olympic swimmers who had "choked" at the Olympic trials or on the Olympic stage supports this idea. When Team Canada's sport psychologist, Hap Davis, and a group of neuroscientists used fMRI to peer inside the brains of these swimmers as they watched videos of their failed races, they found decreased activity in important motor areas of the brain that we use to take action. They also saw a lot of activity in emotional centers of the brain associated with anxiety. But after sport psychologists worked with the athletes to help them have a sense of control over their future swims—to specifically think about what they would do differently with their body in their next big race (a smoother stroke or getting off the blocks faster)—their brain showed less activity in the negative emotion centers and more activity in motor regions the swimmers needed to do their best.[17]

Gaining insight into how we are physically going to alter our failure experiences changes our feelings about whether or not we will be able to succeed the next time around.

Our body and our surroundings affect how we think, reason, act, and experience emotions and feelings much more than we ever imagined. From the way we contort our face, to how we move our hands to gesture, these signals don't just travel in one direction, from the mind to the body. The messages that the body sends to the mind are just as important. In school, in work, and in our relationships, how we act has a big effect on how we think. Whether we are a weekend warrior trying to win a bet with our buddies on the back nine or watching our favorite NBA star dunk the ball during prime time, our brain's ability to simulate the outcomes of our actions and the actions of those around us makes the difference between being mere observers and feeling as if we are part of the team.

Our mind is always working to replay, understand, and predict what will happen in the world around us. So it's easy to see that taking breaks can rejuvenate our brainpower. Indeed, after a week in Italy, roaming through the lush gardens of the villa, ignoring my cell phone and email, and generally not taking part in the hurried pace of urban life, I always feel healthier. Physically and mentally. Up until a few years ago, I didn't give this mental transformation much thought. Perhaps that's not surprising, given that most of my training as a scientist has pushed me to think about the disembodied mind. But my thinking has changed. I no longer see mind and body as separate, and I no longer envision our mind as software running on a body of hardware. Now I realize that my thinking extends beyond the cortex and that I can use my body to ensure that my mind functions at its best.

EPILOGUE

Using Your Body to Change Your Mind

The body has a strong influence on the mind. Whether it's learning in school, creativity in the workplace, or success on the playing field or performance stage, there are countless examples of our physical experiences influencing our thinking. Now we know that the line between mind and body isn't a one-way street. You can use your body, your actions, and your surroundings to change your mind and the minds of those around you.

Here is a recap of the striking power of the body.

Your Body as a Tool to Feel Your Best

✦ Your facial expressions affect how you feel inwardly and even how you react to stress. Put another way, your face does more than express how you are feeling inside; it actually affects how emotions are registered in the brain. When smiling, you feel happier and recover from a painful experience faster. Laughter also seems to offer positive psychological benefits. There really is something to the adage "Grin and bear it."

- Your body has a direct line to your brain and exerts a powerful influence on your mental health and well-being. That's one reason why physical ailments affect your interpretation of psychological pain and rejection. The opposite is true too: people who are depressed tend to experience a higher rate of physical ailments than those who are mentally healthy. Even more striking, taking Tylenol not only helps ease physical pain but can ease the psychological pains of loneliness and rejection.

- How you stand can change your state of mind. Standing in a "power pose," even for just a minute or so, increases feelings of power. On the flip side, your body postures also give clues to others about how you are feeling. A slumped posture tells others you have failed. Your posture can serve as a tool to help you put your best foot forward.

- The body is not a passive device that carries out orders from the brain. Your body sends subtle signals that influence the decisions you make. You like products on the store shelf better when they are easy to grasp and carry. Likewise a lifetime of associations between flexing your arm and gratification means that cradling your grocery store basket in the crook of your arm makes it more likely that you will give in to your desire and buy that candy bar at the checkout counter.

- Knowing that we often understand abstract ideas like morality or luck in physical terms helps us make sense of some seemingly odd behaviors people engage in. We believe that we can wash away our sins—and our good luck too. Just ask an athlete who wouldn't dare wash his pair of lucky socks.

✦ Failing often goes hand in hand with the idea that there is nothing you can do to change your predicament. Research with athletes who are down in the dumps about a failed performance shows that you can change this by using the body. Just thinking about what you are going to do differently the next time around, such as altering your form or technique in a specific way, changes your feelings about whether or not you will succeed.

Your Body as a Tool to Help You Think

✦ Whether it's a wedding toast or a pitch to a client, you can use actions and gestures to help you remember your script because actions help make memories last. Practice picking up a glass as you give a dry run of your toast or incorporate meaningful hand gestures into your speech. That way, when all eyes are on you and you have to remember your lines, your body will be in a position to do some of the remembering for you. Movements can also serve as an effective hook for retrieving thoughts that have slipped your mind.

✦ When stuck on a problem at work, in your personal life, or even on an important test, don't constrain your body, because not moving may constrain your mind too. For example, moving Baoding balls from one hand to the other may help lower your threshold for connecting thoughts that might not normally go together. Certain types of movement can actually help facilitate connections between distant ideas.

✦ Stepping completely away from a problem you are stuck on can increase your probability of success. Walking away from a puzzle or challenge increases the likelihood that new solutions will bubble to the surface of your mind. Just as rebooting your computer can help get rid of temporary bugs, moving away from a problem can help flush out dead-end thinking.

✦ Gestures are not used just to communicate information; they help free up brainpower. Indexing information on our fingertips (say, keeping track of the three points you need to make during your presentation) means you need to hold less in mind. Gestures also serve as a mental scratch pad when you are thinking about complex problems. Using your hands to explain a three-dimensional structure, a molecule or a map, in the space in front of you can help you and others see things more clearly.

✦ Trying to learn a new language as an adult can be difficult, especially when it comes to understanding a native speaker whose sentences and even paragraphs meld together, sounding like one big word. One reason you may have a hard time making sense of foreign pronunciations is that you don't have a lot of practice making the mouth movements needed to produce these sounds yourself. Just as high school football players are good at reading their favorite NFL player's next move on the field because they play the game themselves, producing sounds with your own mouth helps you make sense of the same mouth movements in others.

✦ Need a reason to lace up those sneakers? Fitness is associated with enhanced thinking and reasoning from adolescence through older adulthood and can also enhance creativity. Being able to think about a problem in new and unusual ways is aided by a short bout of aerobic exercise. Next time you are stuck on a problem, get moving.

✦ Meditation can help change your mind. It has been shown to help alleviate anxiety and chronic pain and even reduce symptoms of obsessive-compulsive disorder. But you don't have to meditate for hours to get the benefits. A relatively new meditative practice called integrative body-mind training has been shown to change our brain after only eleven hours of practice. Short bouts of IBMT can also improve self-control—a boon for people trying to control their urge to, say, stop smoking.

✦ Your ability to concentrate is important because it helps you mentally buckle down and stay on a task. Being in nature or even looking at nature can help hone those concentration skills and boost your thinking power.

Using the Body to Help Grow the Mind

✦ Babies learn about the world by physically exploring it. The actions they perform early in life (even before they are one year old) are good predictors of academic achievement later in life. This means that, along with cognitive milestones, parents should work on their children's motor development too.

✦ Babies who spend lots of time in walkers have a harder time walking unaided because they are used to having their weight supported. Even wearing diapers can hinder normal walking. Because motor development is linked to cognitive development, babies should get to run around in their birthday suit as often as practical.

✦ Physical activities can help children learn to read and solve problems. Printing letters helps jump-start areas of the brain needed for reading. Piano practice can enhance finger dexterity, counting competence, and math skills. Helping kids act out the stories they read can enhance their reading comprehension. Playing with blocks can be beneficial for learning, but manipulatives have the most positive learning benefit when they can be directly connected to the content of the problem children are trying to solve. Simply put, how we move can aid how we think. Maria Montessori had it right: the body is an important part of the learning process, if you know how to use it.

Your Body as a Tool to Understand the Experiences of Others

✦ Eager fan behavior (lurching along with the quarterback on TV) may be the result of fans' motor cortex playing along with the athlete they are watching. Having played the game yourself gives you a real edge in predicting whether a throw will be caught or a shot made. Playing experience allows you to act out likely outcomes in your head before they have happened in reality.

✦ We cry while watching sad movies even though we know the story isn't true because we empathize with the characters as if their trials and tribulations were our own. This activates many of the same neural circuits involved in our firsthand experience of pain or sadness. Our neural hardware doesn't always make a distinction between what we see and what we experience ourselves. This is why young doctors have to work very hard to stay emotionally detached from a situation. The merging of self and other happens routinely, and it requires effort and practice to disconnect.

✦ Whether you choose to take on that extra hill on your hike or run is influenced by your body. People who are out of shape view hills as steeper. Being psychologically exhausted has similar effects. The good news is that having a friend close by—or just thinking about a supportive friend—can change how difficult you perceive a physical challenge will be. When accompanied by a friend, you won't think the hill is as steep and you may be more likely to tackle a challenge.

✦ The details of what people are communicating are evident in their words, on their face, and in their hands. Gestures reveal a speaker's feelings about what she is saying, even when she doesn't put those feelings into words. Right-handers tend to gesture about things they like with their right hand, left-handers with their left. Poker players' actions can betray the quality of their cards. Whether you're gesturing when you talk, pushing an offer across the table on a piece of paper, or even shaking hands, what you do with your body reflects what's going on in your mind.

✦ All that time you are spending on your device's keyboard is changing how you think. Your vernacular is linked to how easy it is to type certain words. Because people like to act in ways that are easy, they like words that are easy to type. This is why LOL will likely stick around and why baby names with more right-handed letters have become more popular since the home computer became the norm.

✦ These days, face-to-face meetings are being replaced by video conferencing or telecommuting. Although virtual interactions offer some advantages, being physically close to someone makes you feel more psychologically connected—of one mind—and physical distance encourages psychological distance. So you may want to be less reliant on virtual tools, at least for the most sensitive and important conversations.

Notes

Introduction

1. For an overview, see R. A. Zwaan and D. Pecher, "Revisiting Mental Simulation in Language Comprehension: Six Replication Attempts," (2012), *PLoS ONE* 7: e51382, doi:10.1371/journal.pone.0051382.

Chapter 1

1. Statistics retrieved from http://www.nimh.nih.gov/statistics/1mdd_adult.shtml and http://www.census.gov/popclock/ on August 26, 2013.
2. Gotthold Lessing, as cited in A. J. Fridlund, "Evolution and Facial Action," *Biological Psychology* 32 (1991): 3–100.
3. See P. M. Niedenthal, "Embodying Emotion," *Science* 316 (2007): 1002–5, doi:10.1126/science.1136930, for an overview of findings regarding how facial expressions impact thoughts, feelings, and behavior.
4. T. L. Kraft and S. D. Pressman, "Grin and Bear It: The Influence of Manipulated Facial Expression on the Stress Response," *Psychological Science* 23 (2012): 1372–78. See also A. A. Labroo, A. Mukhopadhyay, and P. Dong, "Not Always the Best Medicine: Why Frequent Smiling Can Reduce Well-Being," *Journal of Experimental Social Psychology* (2014), doi:10.1016/j.jesp.2014.03.001, for evidence that smiling most strongly enhances feelings of well-being when folks believe that people smile *when* they feel good (rather than *to* feel good).

5. J. Cole, *About Face* (Cambridge, MA: MIT Press, 1998), as cited in E. Finzi and E. Wasserman, "Treatment of Depression with Botulinum Toxin A: A Case Series," *Dermatologic Surgery* 32 (2006): 645–50.
6. Aaron T. Beck, Robert A. Steer, Gregory K. Brown, Beck Depression Inventory-II (BDI-11). Pearson.
7. Sample items similar to those found in the Beck Depression Inventory-II (BDI-II). Copyright © 2006, Aaron T. Beck. Reproduced with permission of the publisher, NCS Pearson, Inc. All rights reserved. "Beck Depression Inventory" and "BDI" are trademarks, in the US and/or other countries, of Pearson Education, Inc. or its affiliates(s).
8. M. Perrone, "Botox for Migraines: FDA Approves Botox for Migraine Headaches," *Huffington Post*, October 15, 2010, http://www.huffingtonpost.com/2010/10/18/fda-approves-botox-for-mi_n_766369.html.
9. "Botox Injections Fight Underarm Sweat," WebMD, July 26, 2005, http://www.webmd.com/skin-problems-and-treatments/news/20050726/botox-injections-fight-underarm-sweat.
10. M. B. Lewis and P. J. Bowler, "Botulinum Toxin Cosmetic Therapy Correlates with a More Positive Mood," *Journal of Cosmetic Dermatology* 8 (2009): 24–26, doi:10.1111/j.1473-2165.2009.00419.x.
11. D. A. Havas, A. M. Glenberg, L. A. Gutowski, M. J. Lucarelli, and R. J. Davidson, "Cosmetic Use of Botulinum Toxin-A Affects Processing of Emotional Language," *Psychological Science* (2010), doi:10.1177/0956797610374742.
12. A. Hennenlotter, C. Dresel, F. Castrop, A. O. Ceballos-Baumann, A. M. Wohlschläger, and B. Haslinger, "The Link Between Facial Feedback and Neural Activity within Central Circuitries of Emotion—New Insights from Botulinum Toxin–Induced Denervation of Frown Muscles," *Cerebral Cortex* 19 (2009): 537–42.
13. M. A. Wollmer et al., "Facing Depression with Botulinum Toxin: A Randomized Controlled Trial," *Journal of Psychiatric Research* 46 (2012): 574–81. This study had a rather small sample, and thus a larger trial is warranted. For a relevant discussion of supporting work, see also E. Finzi, *The Face of Emotion: How Botox Affects Our Moods and Relationships* (New York: Macmillan, 2012).
14. W. James, *The Principles of Psychology* (New York: Holt, 1890).
15. C. R. Darwin, *The Expression of Emotion in Man and Animals* (New York: Appleton, 1896).

16. For a review, see N. I. Eisenberger, "Broken Hearts and Broken Bones: A Neural Perspective on the Similarities between Social and Physical Pain," *Current Directions in Psychological Science* 21 (2012): 42–47.
17. S. M. Bigatti et al., "An Examination of the Physical Health, Health Care Use, and Psychological Well-Being of Spouses of People with Fibromyalgia Syndrome," *Health Psychology* 21 (2002): 157–66.
18. K. A. Davies et al., "Insecure Attachment Style Is Associated with Chronic Widespread Pain," *Pain* 143 (2009): 200–205.
19. I. M. Lyons and S. L. Beilock, "Mathematics Anxiety: Separating the Math from the Anxiety," *Cerebral Cortex* (2011), doi:10.1093/cercor/bhr289.
20. C. N. DeWall et al., "Acetaminophen Reduces Social Pain: Behavioral and Neural Evidence," *Psychological Science* 21 (2010): 931–37. This was a research finding and should in no way be taken as a recommendation; you should never take medicine yourself or give children medicine without consulting a physician.

Chapter 2

1. The description of the Breslin family is loosely based on the chronicles of a family dealing with a child with a variety of developmental disorders in the book *I Believe in You: A Mother and Daughter's Special Journey* by Michele Gianetti, R.N. (Mustang, OK: Tate Publishing, 2012).
2. Parts of the story of the Rizzolatti discovery were taken from a *Scientific American* article on mirror neurons: Daniel Lametti, "Mirroring Behavior—How Mirror Neurons Let Us Interact with Others," June 9, 2009, http://www.scientificamerican.com/article.cfm?id=mirroring-behavior. See also "Reflecting on Behavior: Giacomo Rizzolatti Takes Us on a Tour of the Mirror Mechanism," Keynote address, 23rd annual APS meeting, Washington, DC, May 2011.
3. Note that these actions are goal-directed. For an early review of mirror neurons, see V. Gallesse and A. Goldman, "Mirror Neurons and the Simulation Theory of Mind-Reading," *Trends in Cognitive Science* 2 (1998): 493–501. For a more recent treatment, see G. Rizzolatti and C. Sinigaglia, *Mirrors in the Brain: How We Share Our Actions and Emotions* (Oxford: Oxford University Press, 2008). Brain circuits with "mirror properties" are not limited to the premotor cortex but exist in

other brain areas as well. I use the term *mirror neuron* rather loosely when referring to humans to refer to brain systems that are involved in both the perception and execution of actions. So perhaps a better term is *mirror system*.

4. J. Piaget, Piaget's Theory, in P. H. Mussen (ed.), *Carmichael's Manual of Child Psychology* (New York: Wiley, 1970).
5. As described and quoted in Seymour Papert, "Child Psychologist Jean Piaget: He Found the Secrets of Human Learning and Knowledge Hidden Behind the Cute and Seemingly Illogical Notions of Children," *Time*, March 29, 1999.
6. Based on accounts of Piaget on Jeremy Dean's PsyBlog, http://www.spring.org.uk/2008/07/jean-piagets-four-stage-theory-how.php. For further reading, see J. Piaget, *The Origins of Intelligence in Children* (New York: International University Press, 1952); J. Piaget, *The Construction of Reality in the Child* (New York: Basic Books, 1954).
7. Though Piaget is credited with many insights and discoveries regarding child development, other researchers have shown that object permanence does occur earlier in life. See R. Baillargeon, "Object Permanence in 3½- and 4½-Month-Old Infants," *Developmental Psychology* 23 (1987): 655–64; R. Baillargeon, "Infants' Physical World," *Current Directions in Psychological Science* 13 (2004): 89–94.
8. K. E. Adolph et al., "How Do You Learn to Walk? Thousands of Steps and Dozens of Falls per Day," *Psychological Science* 23 (2012): 1387–94.
9. S. Pinker, *The Language Instinct: How the Mind Creates Language* (New York: William Morrow, 1994).
10. For a review, see B. I. Bertenthal, J. J. Campos, and K. C. Barrett, "Self-Produced Locomotion: An Organizer of Emotional, Cognitive, and Social Development in Infancy," in R. Emde and R. Harmon (eds.), *Continuities and Discontinuities in Development* (New York: Plenum, 1984); J. J. Campos et al., "Travel Broadens the Mind," *Infancy* 1 (2000): 149-219.
11. N. Rader, M. Bausano, and J. E. Richards, "On the Nature of the Visual-Cliff-Avoidance Response in Human Infants," *Child Development* 51 (1980): 61–68.
12. See A. Greene, *From First Kicks to First Steps: Nurturing Your Baby's Development from Pregnancy through the First Year of Life* (New York: McGraw-Hill, 2004). See also Greene's *New York Times* blog on the dangers of baby walkers: http://consults.blogs.nytimes.com/2010/02/22/the-dangers-of-baby-walkers/?_r=0.

13. M. Garrett et al., "Locomotor Milestones and Babywalkers: Cross-Sectional Study," *British Medical Journal* 324 (2002): 1494.
14. W. G. Cole, J. M. Lingeman, and K. E. Adolph, "Go Naked: Diapers Affect Infant Walking," *Developmental Science* 15 (2012): 783–90. This study was done with thirteen- and nineteen-month-old infants. Walking impairments due to diapers were equally strong for both ages.
15. J. Herbert et al., "Crawling Is Associated with More Flexible Memory Retrieval by 9-Month-Old Infants," *Developmental Science* 10 (2007): 183–89. Crawling infants exhibited more flexible memories than same-age noncrawling babies. There was no difference in babies' memories for how to use the same toy they had encountered in the past (e.g., what to press on the toy to have it move or make a sound). It was only when babies encountered a new, but similar, toy that crawling infants showed a memory advantage.
16. A. Siegel and R. Burton, "Effects of Babywalkers on Early Locomotor Development in Human Infants," *Journal of Developmental and Behavioral Pediatrics* 20 (1999): 355–61.
17. J. A. Sommerville, A. L. Woodward, and A. N. Needham, "Action Experience Alters 3-Month-Old Infants' Perception of Other's Actions," *Cognition* 96 (2005): 1–11.
18. As told on the dyspraxia.org website, http://www.dyspraxiausa.org/in-the-news/radcliffe-story/.
19. M. H. Bornstein, "Physically Developed and Exploratory Young Infants Contribute to Their Own Long-Term Academic Achievement," *Psychological Science* (2013), doi:10.1177/0956797613479974. This study looked at both infants' motor maturity and exploratory activity as indicators of later academic achievement.
20. K. H. James, "Sensori-Motor Experience Leads to Changes in Visual Processing in the Developing Brain," *Developmental Science* 13 (2010): 279–88. The reader should note that printing practice was somewhat better for letter recognition than naming practice. Though this result fell just short of significance, other work suggests the benefit of printing. See also B. D. McClandiss, "Educational Neuroscience: The Early Years," Proceedings of the National Academy of Sciences (2010), www.pnas.org/cgi/doi/10.1073/pnas.1003431107.
21. F. H. Rauscher, G. L. Shaw, and K. N. Ky, "Music and Spatial Task Performance," *Nature* 365 (1993): 611.

22. Kate Connolly, "Sewage Plant Plays Mozart to Stimulate Microbes," *Guardian*, June 2, 2010, http://www.theguardian.com/world/2010/jun/02/sewage-mozart-germany.
23. Rochelle Jones, "Mozart's Nice but Doesn't Increase IQs," CNN.com, August 25, 1999, http://www.cnn.com/HEALTH/9908/25/mozart.iq/.
24. C. F. Chabris, "Prelude or Requiem for the Mozart Effect?," *Nature* 400 (1990): 826–27.
25. J. Pietschnig et al., "Mozart Effect—Shmozart Effect: A Meta-Analysis," *Intelligence* 38 (2010): 314–23.
26. K. M. Nantais and E. G. Schellenberg, "The Mozart Effect: An Artifact of Preference," *Psychological Science* (1999), doi:10.1111/1467-9280.00170.
27. See http://mathcounts.org/.
28. See http://thefundsa.blogspot.com/2012_03_01_archive.html for details.
29. M. Andres et al., "Contribution of Hand Motor Circuits to Counting," *Journal of Cognitive Neuroscience* 19 (2007): 1–14; M. Andres et al., "Actions, Words, and Numbers: A Motor Contribution to Semantic Processing?," *Current Directions in Psychological Science* 17 (2008): 313–17; M. Andres et al., "Common Substrate for Mental Arithmetic and Finger Representation in the Parietal Cortex," *Neuroimage* 62 (2012): 1520–28.
30. For example, see E. Mayer, "A Pure Case of Gerstmann Syndrome with a Subangular Lesion," *Brain* 22 (1999): 1107–20.
31. L. R. Moo et al., "Interlocking Finger Test: A Bedside Screen for Parietal Lobe Dysfunction," *Journal of Neurology, Neurosurgery, & Psychiatry* 74 (2003): 530–32.
32. S. Di Luca, A. Granà, C. Semenza, X. Seron, and M. Pesenti, "Finger-Digit Compatibility in Arabic Numeral Processing," *Quarterly Journal of Experimental Psychology* 59 (2006): 1648–63. See also M. H. Fischer, "Finger Counting Habits Modulate Spatial-Numerical Associations, *Cortex* 44 (2008): 386–92; F. Domahs et al., "Embodied Numerosity: Implicit Hand-Based Representations Influence Symbolic Number Processing across Cultures," *Cognition* 116 (2010): 251–66.
33. B. Butterworth, *The Mathematical Brain* (London: Macmillan, 1999), quoted in M. Noel, "Finger Gnosia: A Predictor of Numerical Abili-

ties in Children?," *Child Neuropsychology* 11 (2005): 413–30. Others have argued that the relations found between math and fingers could be due to the fact that the brain areas underlying these two abilities are merely close together rather than the same.

34. For a review of the literature regarding these findings, see M. P. Noël, "Finger Gnosia: A Predictor of Numerical Abilities in Children?," *Child Neuropsychology* 11 (2005): 413–30; M. Gracia-Bafalluy and M. P. Noël, "Does Finger Training Increase Young Children's Numerical Performance?," *Cortex* 44 (2008): 368–75; I. Imbo et al., "Passive Hand Movements Disrupt Adults' Counting Strategies," *Frontiers in Psychology* 2 (2011): 1–5; M. Penner-Wilger and M. L. Anderson, "The Relation Between Finger Gnosis and Mathematical Ability: Why Redeployment of Neural Circuits Best Explains the Finding," *Frontiers in Psychology* (2013), doi:10.3389/fpsyg.2013.00877.
35. See B. Butterworth, *The Mathematical Brain* (London: Macmillan, 1999), quoted in M. P. Noël, "Finger Gnosia: A Predictor of Numerical Abilities in Children?" *Child Neuropsychology* 11 (2005): 413–30.
36. M. Gracia-Bafalluy and M. P. Noël, "Does Finger Training Increase Young Children's Numerical Performance?" *Cortex* 44 (2008): 368–75. See also J. P. Fischer, "Numerical Performance Increased by Finger Training: A Fallacy Due to Regression toward the Mean?" *Cortex* 46 (2010): 272–73. M. Gracia-Bafalluy et al., "Consequences of Playing a Musical Instrument on Finger Gnosia and Number Skills in Children" (2007, June), paper presented at Numbers, Fingers, and the Brain symposium at the Belgian Association for Psychological Sciences Annual Meeting at the Université Catholique de Louvain. A. B. Graziano et al., "Enhanced Learning of Proportional Math through Music Training and Spatial-Temporal Training," *Neurological Research* 21 (1999): 139–52. V. J. Schmithorst and S. K. Holland, "The Effect of Musical Training on the Neural Correlates of Math Processing: A Functional Magnetic Resonance Imaging Study in Humans," *Neuroscience Letters* 354 (2004): 193–96. K. Vaughn, "Music and Mathematics: Modest Support for the Oft-Claimed Relationship," *Journal of Aesthetic Education* 34 (2000): 149–66. The reader should note that more work on the relation between music and math is needed.

Chapter 3

1. See http://www.eng.cam.ac.uk/news/daniel-wolpert-real-reason-brains. For more on the sea squirt, see R. Llinas, *The I of the Vortex* (Cambridge, MA: MIT Press, 2001).
2. O. Hauk, I. Johnsrude, and F. Pulvermüller, "Somatotopic Representation of Action Words in Human Motor and Premotor Cortex," *Neuron* 41 (2004): 301–7.
3. A. M. Glenberg et al., "Activity and Imagined Activity Can Enhance Young Children's Reading Comprehension," *Journal of Educational Psychology* 96 (2004): 424–36.
4. See J. P. Gee, "Reading as a Situated Language: A Sociocognitive Perspective," in R. B. Ruddell et al. (eds.), *Theoretical Models and Processes of Reading,* 6th edition, Newark, DE: International Reading Association (2013).
5. For a review, see G. Lakoff and R. E. Nunez, *Where Mathematics Comes From* (New York: Basic Books, 2000).
6. For a review, see J. P. Gee, "Reading as a Situated Language: A Sociocognitive Perspective," in R. B. Ruddell et al. (eds.), *Theoretical Models and Processes of Reading,* 6th edition, Newark, DE: International Reading Association (2013).
7. A. M. Glenberg, B. Jaworski, M. Rischal, and J. R. Levin, "What Brains Are For: Action, Meaning, and Reading Comprehension," in D. McNamara (ed.), *Reading Comprehension Strategies: Theories, Interventions, and Technologies,* Mahwah, NJ: Lawrence Erlbaum (2007).
8. Kyle Spencer, "With Building Blocks, Educators Going Back to Basics," *New York Times,* November 27, 2011.
9. National Council of Teachers of Mathematics, *Principles and Standards for School Mathematics,* Reston, VA: NCTM (2000). See also Jean M. Shaw, "Manipulatives Enhance the Learning of Mathematics" (2002), http://www.eduplace.com/state/author/shaw.pdf.
10. M. Montessori, *The Absorbent Mind* (C. A. Claremong, trans.) (New York: Holt, 1967), cited in A. M. Glenberg, B. Jaworski, M. Rischal, and J. R. Levin, "What Brains Are For: Action, Meaning, and Reading Comprehension," in D. McNamara (ed.), *Reading Comprehension Strategies: Theories, Interventions, and Technologies,* Mahwah, NJ: Lawrence Erlbaum (2007).

11. K. Schaffer, E. Stern, and S. Kim, *Math Dance with Dr. Schaffer and Mr. Stern*, Santa Cruz, CA: MoveSpeakSpin (2001).
12. As quoted in "Do the Math Dance," *Science Daily*, May 1 (2008), http://old.sciencedaily.com/videos/2008/0503-do_the_math_dance.htm.
13. As described in Lisa Traiger "1 + 1 = Pas de deux," *Dance Teacher Magazine*, March 15 (2010), http://www.dance-teacher.com/2010/03/1-1-pas-de-deux/.
14. "Tom Daley Fears He May Grow Too Tall to Dive," BBC News, October 14 (2010), http://www.bbc.co.uk/newsbeat/11541138.
15. Malcolm Folley, "This Medal's for You Dad! Poster Boy Daley Delivers Bronze in Diving Thriller," *Daily Mail* (UK), August 11 (2012), http://www.dailymail.co.uk/sport/olympics/article-2187152/London-Olympics-2012-Tom-Daley-wins-diving-bronze.html.
16. For a review, C. Kontra, S. Goldin-Meadow, and S. L. Beilock, "Embodied Learning across the Lifespan," *Topics in Cognitive Science* 4 (2012): 731–39. In this work, we specifically talk about our experiments exploring the concept of angular momentum and the related concept of torque.

Chapter 4

1. See M. Gick and K. Holyoak, "Analogical Problem Solving," *Cognitive Psychology* 12 (1980), 306–56; M. Gick and K. Holyoak, "Scheme Induction and Analogical Transfer," *Cognitive Psychology* 15 (1983): 1–38; K. Duncker, "On Problem Solving," *Psychological Monographs* 58 (1945), 270.
2. L. E. Thomas and A. Lleras, "Moving Eyes and Moving Thought: On the Spatial Compatibility between Eye Movements and Cognition," *Psychonomic Bulletin & Review* 14 (2007): 663–68. See L. E. Thomas and A. Lleras, "Covert Shifts of Attention Function as an Implicit Aid to Insight," *Cognition* 111 (2009): 168–74, for evidence that simply shifting attention (an act that often precedes actual eye movements) to the problem solution produces similar results.
3. E. R. Grant and M. J. Spivey, "Eye Movements and Problem Solving: Guiding Attention Guides Thought," *Psychological Science* 14 (2003): 462–66.

4. D. Kirsh, "Creative Cognition in Choreography," paper presented at Proceedings of 2nd International Conference on Computational Creativity. Mexico City, Mexico (2011, April 27-29).
5. A. K.-Y. Leung et al., "Embodied Metaphors and Creative Acts," *Psychological Science* 23 (2012), 502–9. A reader interested in these findings might also read G. Francis, "The Frequency of Excessive Success," *Psychonomic Bulletin & Review* (in press). For more effects of movement on problem solving and creativity, see K. Werner and M. Raab, "Moving to Solution: Effects of Movement Priming on Problem Solving," *Experimental Psychology* 60 (2013): 403–9, and M. Oppezzo and D. Schwartz, "Give Your Ideas Some Legs: The Positive Effect of Walking on Creative Thinking," *Journal of Experimental Psychology: Learning, Memory, and Cognition* 40, 1142–52.
6. Quoted in Nadia Goodman, "3 Postures to Boost Productivity Now," Entrepreneur.com, May 31, 2012.
7. D. R. Carney, A. J. Cuddy, and A. J. Yap, "Power Posing: Brief Nonverbal Displays Affect Neuroendocrine Levels and Risk Tolerance," *Psychological Science* 10 (2010): 1363–68. See also Nadia Goodman, "3 Postures to Boost Productivity Now," Entrepreneur.com, May 31, 2012. See also J. Cesario and M. M. McDonald, "Bodies in Context: Power Poses as a Computation of Action Possibilities," *Social Cognition* 31 (2013): 260–74. Note that more information about the body position tips mentioned can be found in these citations, including information about how the context one is in can modulate the effect that power poses have on perceptions of power. Also, note that there is some controversy regarding whether power poses change perceptions of power via hormones, such as testosterone or other mechanisms. See E. Ranehill et al., "A Reassessment of Power Posing and Risk Tolerance: No Effect in a Large Sample of Men and Women," *Psychological Science* (in press).
8. See S. E. Gaither and S. R. Sommers, "Having an Out Group Roommate Shapes Whites' Behavior in Subsequent Diverse Settings," *Journal of Experimental Social Psychology* 49 (2013): 272–76, doi:10.1016/j.jesp.2012.10.020; J. N. Shelton, J. A. Richeson, and J. Salvatore, "Expecting to Be the Target of Prejudice: Implications for Interethnic Interactions," *Personality and Social Psychology Bulletin* 31 (2005): 1189–202.
9. H. Aviezer et al., "Body Cues, not Facial Expressions, Discriminate between Intense Positive and Negative Emotions," *Science* 338 (2012), doi:10.1126/science.1224313.

10. A. J. Yap et al., "The Ergonomics of Dishonesty: The Effect of Incidental Posture on Stealing, Cheating, and Traffic Violations," *Psychological Science* (2013), doi:10.1177/0956797613492425.
11. T. Noice and H. Noice, *The Nature of Expertise in Professional Acting: A Cognitive View*, Mahwah, NJ: Lawrence Erlbaum (1997). See also T. Noice and H. Noice, "What Studies of Actors and Acting Can Tell Us about Memory and Cognitive Functioning," *Current Directions in Psychological Science* 15 (2006): 14–18.
12. H. Noice, T. Noice, and C. Kennedy, "The Contribution of Movement on the Recall of Complex Material," *Memory* 8 (2000): 353–63.
13. Example taken from T. Noice and H. Noice, "What Studies of Actors and Acting Can Tell Us about Memory and Cognitive Functioning," *Current Directions in Psychological Science* 15 (2006): 14–18.
14. H. Noice, T. Noice, and G. Staines, "A Short-Term Intervention to Enhance Cognitive and Affective Functioning in Older Adults," *Journal of Aging and Health* 16 (2004): 1–24.
15. For a review, see H. L. Roediger, III and J. D. Karpicke, "The Power of Testing Memory: Basic Research and Implications for Educational Practice," *Perspectives on Psychological Science* 1 (2006): 181–210.
16. For a review, see P. F. Delaney et al., "Spacing and the Testing Effects: A Deeply Critical, Lengthy, and at Times Discursive Review of the Literature," *Psychology of Learning and Motivation* 53 (2010): 63–147.
17. For a review, see J. Dunlosky et al., "Improving Students' Learning with Effective Learning Techniques: Promising Direction from Cognitive and Educational Psychology," *Psychological Science in the Public Interest* 14 (2013): 4–58.

Chapter 5

1. For details, see "McCain Puts Obama on the Spot in Final Debate," CNNPolitics.com, October 16, 2008, http://www.cnn.com/2008/POLITICS/10/15/presidential.debate/.
2. See Larry Vellequette and Tom Troy, "'Joe the Plumber' Isn't Licensed," *Toledo Blade*, October 16, 2008, http://www.toledoblade.com/Politics/2008/10/16/Joe-the-plumber-isn-t-licensed.html.
3. See a video of the third 2008 Presidential Debate at http://www.youtube.com/watch?v=DvdfO0lq4rQ.

4. See D. Casasanto and K. Jasmin, "Good and Bad in the Hands of Politicians" (2010), *PLoS ONE* 5, e11805, doi:10.1371/journal.pone.0011805; D. Casasanto and E. G. Chrysikou, "When Left Is 'Right': Motor Fluency Shapes Abstract Concepts," *Psychological Science* 22 (2011), 419–22; D. Casasanto, "Different Bodies, Different Minds: The Body-Specificity of Language and Thought," *Current Directions in Psychological Science* 20 (2011): 378–83.
5. D. Casasanto, "Embodiment of Abstract Concepts: Good and Bad in Right- and Left-Handers," *Journal of Experimental Psychology: General* 138 (2009): 351–67.
6. D. Casasanto and K. Jasmin, "Good and Bad in the Hands of Politicians" (2010), *PLoS ONE* 5: e11805, doi:10.1371/journal.pone.0011805.
7. Abha Bhattarai, "Executives Turn to Body Language for an Edge," *Washington Post*, February 24, 2013.
8. Personal communication with Daniel Casasanto, September 26, 2011.
9. M. L. Slepian et al., "Quality of Professional Players' Poker Hands Is Perceived Accurately from Arm Motions," *Psychological Science* (2013), doi:10.1177/0956797613487384 (as judged at the 2009 World Series of Poker tournament).
10. S. W. Cook and S. Goldin-Meadow, "The Role of Gesture in Learning: Do Children Use Their Hands to Change Their Minds?," *Journal of Cognition and Development* 7 (2006): 211–32; S. Goldin-Meadow, S. W. Cook, and Z. A. Mitchell, "Gesturing Gives Children New Ideas about Math," *Psychological Science* 20 (2009): 267–72.
11. A. B. Hostettler and M. W. Alibali, "Visible Embodiment: Gestures as Simulated Action," *Psychonomic Bulletin and Review* 15 (2008): 495–514. For an interesting discussion of these issues, see R. Stevens, "The Missing Bodies of Mathematical Thinking and Learning Have Been Found," *Journal of Learning Sciences* 21 (2012): 337–46.
12. See S. Ehrlich, S. C. Levine, and S. Goldin-Meadow, "The Importance of Gesture in Children's Spatial Reasoning," *Developmental Psychology* 42 (2006): 1259–68; S. Goldin-Meadow, S. L. Levine, E. Zinchenko, T. K. Yip, N. Hemani, and L. Factor, "Doing Gesture Promotes Learning a Mental Transformation Task Better Than Seeing Gesture," *Developmental Science* 15 (2012): 876–84.
13. For a review, see S. Goldin-Meadow and S. L. Beilock, "Action's Influence on Thought: The Case of Gesture," *Perspectives on Psychological Science* 5 (2010): 664–74.

14. See webpage of chemistry professor François G. Amar, http://chemistry.umeche.maine.edu/~amar/amar.html.
15. S. W. Cook, T. K. Yip, and S. Goldin-Meadow, "Gesturing Makes Memories that Last," *Journal of Memory and Language* 63 (2010): 465–75.
16. For a review of the QWERTY keyboard, see http://en.wikipedia.org/wiki/QWERTY. See also P. David, "Clio and the Economics of QWERTY," *American Economic Review* 75 (1985): 332–37.
17. Jared Diamond, "The Curse of QWERTY," *Discover Magazine*, April 1, 1997, http://discovermagazine.com/1997/apr/thecurseofqwerty1099/#.UkswzoYqjzY.
18. See entry on Typing: http://en.wikipedia.org/wiki/Typing#Alphanumeric_entry. Several other typing competitions have come on the scene in the past decade (e.g., The Ultimate Typing Championship), and many new records have been set (e.g., fastest typing of numbers).
19. K. Jasmin and D. Casasanto, "The QWERTY Effect: How Typing Shapes the Meaning of Words," *Psychonomic Bulletin & Review* (2012), doi:10.3758/s13423-012-0229-7.
20. D. Casasanto et al., "The QWERTY Effect: How Multiple Language Production Shapes Our Lexicons," paper submitted to the 6th Conference of the International Society for Gesture Studies. San Diego, CA. July 8–11, 2014.
21. S. L. Beilock and L. E. Holt, "Embodied Preference Judgments: Can Likeability Be Driven by the Motor System?" *Psychological Science* 18 (2007): 51–57. In the actual experiment, the letter pairs that were easier to type were sometimes presented on the left side of the computer screen and sometimes on the right side.
22. S. Topolinski, "I 5683 You: Dialing Phone Numbers on Cell Phones Activates Key-Concordant Concepts," *Psychological Science* (2011), doi:10.1177/0956797610397668.
23. For an overview of the end-state-comfort idea, see D. A. Rosenbaum et al., "Plans for Grasping Objects," in M. L. Latash and F. Lestienne (eds.), *Motor Control & Learning* (New York: Springer, 2006); W. Zhang, and D. A. Rosenbaum, "Experimental Brain Research. Planning for Manual Positioning: The End-State Comfort Effect for Manual Abduction-Adduction," *Experimental Brain Research* 184 (2008): 383–89; D. A. Rosenbaum, K. M. Chapman, C. J. Coelho, L. Gong, and B. E. Studenka, "Choosing Actions," *Frontiers in Psychology* 4 (2013), doi:10.3389/fpsyg.2013.00273.

24. D. J. Weiss et al., "Monkey See, Monkey Plan, Monkey Do: The End-State Comfort Effect in Cotton-Top Tamarins (*Saguinus oedipus*)," *Psychological Science* 18 (2007): 1063–68.
25. R. Ping, S. Dhillon, and S. L. Beilock, "Reach for What You Like: The Body's Role in Shaping Preferences," *Emotion Review* 1 (2009): 140–50.
26. Starting in 2008, see the Coca-Cola Company news release: http://web.archive.org/web/20110101092057/ and http://www.thecoca-colacompany.com/presscenter/nr_20080613_2l_contour.html. See also "Flat Sales Send Pepsi to No. 3, behind Coke, Diet Coke," *Daily Record* (MD), March 17, 2011, http://thedailyrecord.com/2011/03/17/flat-sales-send-pepsi-to-no-3-behind-coke-diet-coke/.
27. Chuck Salter, "Mauro Porcini Leaves 3M for Pepsico," *Fast Company Magazine*, September 2012, http://www.fastcompany.com/3000005/mauro-porcini-leaves-3m-pepsico.
28. B. Van Den Bergh et al., "Embodied Myopia," *Journal of Marketing Research* 48 (2011): 1033–44. The basket condition was a small sample in the naturalistic supermarket observation study; most people opted for carts.

Chapter 6

1. For overviews, see http://en.wikipedia.org/wiki/Franz_Joseph_Gall; http://www.victorianweb.org/science/phrenology; http://www.historyofphrenology.org.uk/overview.htm; D. P. Schultz and S. E. Schultz, *A Modern History of Psychology*, 10th edition (Belmont, CA: Thomson Wadsworth, 2013).
2. For a review, see K. M. Galotti, *Cognitive Psychology: In and Out of the Laboratory*, 3rd edition (Belmont, CA: Thomson Wadsworth, 2004). See also P. C. Wason, "Reasoning about a Rule," *Quarterly Journal of Experimental Psychology* 20 (1968): 273–81.
3. O. Hauk, I. Johnsrude, and F. Pulvermüller, "Somatotopic Representation of Action Words in Human Motor and Premotor Cortex," *Neuron* 41 (2004): 301–7.
4. A. M. Glenberg and M. P. Kaschak, "Grounding Language in Action," *Psychonomic Bulletin & Review* 9 (2002): 558–65.

5. R. A. Zwaan and L. J. Taylor, "Seeing, Acting, Understanding: Motor Resonance in Language Comprehension," *Journal of Experimental Psychology: General* 135 (2006): 1–11. People were quicker at reading verbs that matched the action they were preforming (e.g., turning a knob counterclockwise led to faster reading of the verb in the sentence "The marathon runner eagerly opened the water bottle").
6. This patient's story is adapted from reports of patients presenting with motor neuron disease symptoms. See T. H. Bak and J. R. Hodges, "The Effects of Motor Neurone Disease on Language: Further Evidence," *Brain and Language* 89 (2004): 354–61. See also http://www.ninds.nih.gov/disorders/motor_neuron_diseases/detail_motor_neuron_diseases.htm.
7. See the website of the International Alliance of ALS/MND Associations, http://www.alsmndalliance.org/.
8. T. H. Bak et al., "Selective Impairment of Verb Processing Associated with Pathological Changes in Brodmann Areas 44 and 45 in the Motor Neurone Disease-Dementia-Aphasia Syndrome," *Brain* 124 (2001): 103–20.
9. We have a similar map for sensory areas of the brain that represents the sensory information coming from various body parts.
10. W. Penfield and T. Rasmussen, *The Cerebral Cortex of Man* (New York: Macmillan, 1950).
11. Figure adapted from O. Hauk, I. Johnsrude, and F. Pulvermüller, "Somatotopic Representation of Action Words in Human Motor and Premotor Cortex," *Neuron* 41 (2004): 301–7. Reprinted with permission.
12. V. S. Ramachandran, "Phantom Limbs, Neglect Syndromes, Repressed Memories, and Freudian Psychology," *International Review of Neurobiology* 37 (1994): 291–333; V. S. Ramachandran and S. Blakeslee, *Phantoms in the Brain: Human Nature and the Architecture of the Mind* (London: Fourth Estate, 1998); V. S. Ramachandran, *Phantoms in the Brain: Probing the Mysteries of the Human Mind* (New York: Harper, 1999). Feet, toes, and genitals are mapped next to each other in the brain's body map of the sensory cortex.
13. See P. M. Di Noto, "The Hermunculus: What is Known about the Representation of the Female Body in the Brain?" *Cerebral Cortex* (2012), doi:10.1093/cercor/bhs005.
14. For a review, see F. Pulvermüller and M. L. Berthier, "Aphasia Ther-

apy on a Neuroscience Basis," *Aphasiology* 22 (2008): 563–99. For an alternate view, an interested reader can see A. Caramazza et al., "Embodied Cognition and Mirror Neurons: A Critical Assessment," *Annual Review of Neuroscience* 37 (2014): 1–15.

15. L. Wittgenstein, *Philosophical Investigations* (Oxford: Blackwell, 1953), quoted in F. Pulvermüller and M. L. Berthier, "Aphasia Therapy on a Neuroscience Basis," *Aphasiology* 22 (2008): 563–99.
16. See World Health Organization website, http://www.who.int/topics/cerebrovascular_accident/en/; also the CDC, http://www.cdc.gov/stroke/ and http://www.strokecenter.org/patients/about-stroke/stroke-statistics/.
17. See F. Pulvermüller and M. L. Berthier, "Aphasia Therapy on a Neuroscience Basis," *Aphasiology* 22 (2008): 563–99. There are other important principles of aphasia therapy, such as training schedule. See also F. Pulvermüller et al., "Therapy-Related Reorganization of Language in Both Hemispheres of Patients with Chronic Aphasia," *Neuroimage* 28 (2005): 481–89.
18. P. Adank, P. Hagoort, and H. Bekkering, "Imitation Improves Language Comprehension," *Psychological Science* (2010), doi:10.1177/0956797610389192.
19. Example taken from V. Gallese and G. Lakoff, "The Brain's Concepts: The Role of the Sensory-Motor System in Conceptual Knowledge," *Cognitive Neuropsychology* 22 (2005): 455–79.
20. For a review, see A. M. Glenberg, M. Sato, L. Cattaneo, L. Riggio, D. Palombo, and G. Buccino, "Processing Abstract Language Modulates Motor System Activity," *Quarterly Journal of Experimental Psychology* 61 (2008), 905–19. In many of these studies, participants pushed a button close to them and one far away rather than pulled or pushed a lever. See also L. W. Barsalou, "Grounded Cognition," *Annual Review of Psychology* 59 (2008): 617–45.
21. L. Boroditsky and M. Ramscar, "The Roles of Body and Mind in Abstract Thought," *Psychological Science* 13 (2002): 185–89.
22. K. L. Miles, L. K. Nind, and N. Macrae, "Moving Through Time," *Psychological Science* 21 (2010), doi:10.1177/0956797609359333.
23. S. L. Beilock, I. M. Lyons, A. Mattarella-Micke, H. C. Nusbaum, and S. L. Small, "Sports Experience Changes the Neural Processing of Action Language," *Proceedings of the National Academy of Sciences,*

USA 105 (2008): 13269–73. The players listened to the reading of sentences about hockey games rather than an actual radio broadcast.

24. J. Atwood, *Capoeira: A Martial Art and a Cultural Tradition* (New York: Rosen, 1999).
25. B. Calvo-Merino et al., "Action Observation and Acquired Motor Skills: An fMRI Study with Expert Dancers," *Cerebral Cortex* 15 (2005): 1243–49; B. Calvo-Merino et al., "Seeing or Doing? Influence of Visual and Motor Familiarity in Action Observation," *Current Biology* 16 (2006), 1905–10.
26. S. M. Aglioti et al., "Action Anticipation and Motor Resonance in Elite Basketball Players," *Nature Neuroscience* (2008), doi:10.1038/nn.2182. MEPs, or motor-evoked potentials, were used to measure corticospinal activation during observation of basketball shots.
27. B. Abernethy and D. G. Russell, "The Relationship between Expertise and Visual Search Strategy in a Racquet Sport," *Human Movement Science* 6 (1987): 283–319.
28. For an overview of forward models, see K. Yarrow, P. Brown, and J. W. Krakauer, "Inside the Brain of an Elite Athlete: The Neural Processes that Support High Achievement in Sports," *Nature Reviews Neuroscience* 10 (2009): 585–96. For a review of how humans are able to predict and understand the actions of others, see N. Sebanz and G. Knoblich, "Prediction in Joint Action: What, When, and Where," *Topics in Cognitive Science* 1 (2009): 353–67.
29. For more on the idea that athletes don't always have conscious access to what they are doing, see S. L. Beilock, *Choke: What the Secrets of the Brain Reveal about Getting It Right When You Have To* (New York: Free Press, 2010).

Chapter 7

1. The reader should note that the idea of shared neural representations does not necessarily imply mirror neurons. Moreover, it is likely that the brain circuits involved in empathy have more general functions than simply detecting physical pain and social distress (in one's self or others). Specifically, any potential threat likely elicits activation in these neural regions. For more information, see J. Decety, "The Neu-

roevolution of Empathy and Caring for Others: Why It Matters for Morality," in J. Decety and Y. Christen (eds.), *New Frontiers in Social Neuroscience, Research and Perspectives in Neurosciences* 21 (2014), doi:10.1007/978-3-319-02904-7_8.

2. P. Ruby and J. Decety, "How Would You Feel versus How Do You Think She Would Feel? A Neuroimaging Study of Perspective Taking with Social Emotions," *Journal of Cognitive Neuroscience* 16 (2004): 988–99.
3. B. Wicker, C. Keysers, J. Plailly, J.-P. Royet, V. Gallese, and G. Rizzolatti, "Both of Us Disgusted in My Insula: The Common Neural Basis of Seeing and Feeling Disgust," *Neuron* 40 (2003): 655–64. For a review, see P. M. Niedenthal et al., "Embodiment in Attitudes, Social Perception, and Emotion," *Personality and Social Psychology Review* 9 (2005): 184–211.
4. For a review, see J. Decety and M. Meyer, "From Emotion Resonance to Empathic Understanding: A Social Developmental Neuroscience Account," *Development and Psychopathology* 20 (2008): 1053–80.
5. T. Field, B. Healy, S. Goldstein, and M. Guthertz, "Behavior-State Matching and Synchrony in Mother-Infant Interactions of Non-Depressed versus Depressed Dyads," *Developmental Psychology* 26 (1990): 7–14.
6. For a review of Niedenthal and colleagues' findings, see P. M. Niedenthal et al., "When Did Her Smile Drop? Facial Mimicry and the Influences of Emotional State on the Detection of Change in Emotional Expression," *Cognition and Emotion* 15 (2001): 853–64; P. M. Niedenthal, "Embodying Emotion," *Science* 316 (2007): 1002–5.
7. R. B. Zajonc, P. K. Adelmann, S. T. Murphy, and P. M. Niedenthal, "Convergence in the Physical Appearance of Spouses," *Motivation and Emotion* 11 (1987): 335–46.
8. Specifically, brain areas including the somatosensory cortex, anterior insula, dorsal anterior cingulate cortex, anterior medial cingulate cortex, and periaqueductal gray. See J. Decety et al., "Physicians Down-Regulate Their Pain Empathy Response: An Event-Related Brain Potential Study," *Neuroimage* (2010), doi:10.1016/j.neuroimage.2010.01.025.
9. Specifically prefrontal regions underlying executive functions and self-regulation (dorsolateral and medial prefrontal cortex). Y. Cheng et al., "Expertise Modulates the Perception of Pain in Others," *Current Biology* 17 (2007): 1708–13.
10. See G. Ramirez and S. L. Beilock, "Writing about Testing Worries

Boosts Exam Performance in the Classroom," *Science* 331 (2011), 211–13; K. Kircanski et al., "Feelings into Words: Contributions of Language to Exposure Therapy," *Psychological Science* (2012), doi:10.1177/0956797612443830.

11. R. Saxe, "Theory of Mind (Neural Basis)," Vol. 2, 401–10, in *Encyclopedia of Consciousness* (Oxford: Academic Press, 2009). See also J. Decety and M. Meyer, "From Emotion Resonance to Empathetic Understanding: A Social Developmental Neuroscience Account," *Development and Psychopathology* 20 (2008): 1053–80.
12. N. Barnea-Goraly, H. Kwon, V. Menon, S. Eliez, L. Lotspeich, and A. L. Reiss, "White Matter Structure in Autism: Preliminary Evidence from Diffusion Tensor Imaging," *Biological Psychiatry* 55 (2004): 323–26.
13. Autism spectrum disorders (2013), CDC, December 26, http://www.cdc.gov/ncbddd/autism/data.html.
14. For a review of this viewpoint, see V. S. Ramachandran and L. M. Oberman, "Broken Mirrors: A Theory of Autism," *Scientific American*, 62–69 (2006, November). For an argument against a dysfunctional mirror neuron system being the sole driving force behind autism, see V. Southgate and A. Hamilton, "Unbroken Mirrors: Challenging a Theory of Autism" (2008), doi:10.1016/j.tics.2008.03.005.
15. For details, see V. S. Ramachandran and L. M. Oberman, "Broken Mirrors: A Theory of Autism," *Scientific American* (2006, November): 62–69. See also L. Oberman et al., "Modulation of Mu Suppression in Children with Autism Spectrum Disorders in Response to Familiar or Unfamiliar Stimuli: The Mirror Neuron Hypothesis," *Neuropsychologia* 46 (2008): 1558–65. This study revealed that mu suppression was sensitive to the degree of familiarity of the person performing the action participants observed. But also see Y. Fan et al., "Unbroken Mirror Neurons in Autism Spectrum Disorders," *Journal of Child Psychology and Psychiatry* 51 (2010): 981–88. Though they did not find evidence of mu suppression in ASD participants, they did find that more mu suppression to action observation was associated with more communication competence.
16. J. Pineda et al., "Positive Behavioral and Electrophysiological Changes Following Neurofeedback Training in Children with Autism," *Research in Autism Spectrum Disorders* 2 (2008): 557–81. The sample size of this study was small; thus, though intriguing, this work should be considered exploratory.

17. A. Hamilton, "Reflecting on the Mirror Neuron System in Autism: A Systematic Review of Current Theories," *Developmental Cognitive Neuroscience* 3 (2013): 91–105.
18. See D. J. Greene, "Atypical Neural Networks for Social Orienting in Autism Spectrum Disorders," *NeuroImage* 56 (2011): 354–62; J. D. Rudie, "Reduced Functional Integration and Segregation of Distributed Neural Systems Underlying Social and Emotional Information Processing in Autism Spectrum Disorders," *Cerebral Cortex* 22 (2012): 1025–37.

Chapter 8

1. For more on this viewpoint, see http://www.washingtontimes.com/news/2011/feb/14/wetzstein-tips-on-how-to-love-your-child/.
2. See H. F. Harlow, "The Nature of Love," *American Psychologist* 13 (1958): 673–98; D. Blum, *Love at Goon Park: Harry Harlow and the Science of Affection* (New York: Basic Books, 2011). Names of monkeys have been changed.
3. T. K Inagaki and N. I. Eisenberger, "Shared Neural Mechanisms Underlying Social Warmth and Physical Warmth," *Psychological Science* (2013), doi:10.1177/0956797613492773.
4. Y. Kang et al., "Physical Temperature Effects on Trust Behavior: The Role of the Insula," *Social Cognitive and Affective Neuroscience* 6 (2011): 507–15. The researchers conclude that cold activates insula, and that this activation spreads into areas of the anterior insula, which affects subsequent trust decisions.
5. C. B. Zhong and G. J. Leonardelli, "Cold and Lonely: Does Social Exclusion Literally Feel Cold?" *Psychological Science* 19 (2008): 838–42.
6. H. F. Harlow, "The Nature of Love." Address of the president at the sixty-sixth annual convention of the American Psychological Association, Washington D. C. *American Psychologist* 13 (1958): 573–685.
7. Though there does seem to be an overlap in our psychological and physical gauges of temperature, more work is needed to tease apart exactly what that overlap is. For a discussion as it relates to loneliness and warm baths, see http://traitstate.wordpress.com/2012/09/20/whats-the-first-rule-about-john-barghs-data/.
8. The reader should note that more work is needed to assess the value

of these specific activities. In terms of romantic movies, this is true as long as you associate romance movies with psychological warmth; see J. Hong and Y. Sun, "Warm It Up with Love: The Effect of Physical Coldness on Liking of Romance Movies," *Journal of Consumer Research* 39 (2011): 293–306.

9. See N. I. Eisenberger, "The Pain of Social Disconnection: Examining the Shared Neural Underpinnings of Physical and Social Pain," *Nature Reviews Neuroscience* 13 (2012): 421–34; E. Kross et al., "Social Rejection Shares Somatosensory Representations with Physical Pain," *Proceedings of the National Academy of Sciences of the United States of America* 108 (2011): 6270–75. But also see S. Cacioppo et al., "A Quantitative Meta-Analysis of Functional Imaging Studies of Social Rejection," *Scientific Reports* 3 (2013): 2027, doi:10.1038/srep02027 for a discussion of the idea that the neural correlates of social pain are more complex than a simple reliance on the pain matrix.
10. R. Sapolsky, "This is Your Brain on Metaphors," *New York Times*, November 14, 2010, http://opinionator.blogs.nytimes.com/2010/11/14/this-is-your-brain-on-metaphors/?_r=0.
11. For more on this view, see N. I. Eisenberger and M. D. Lieberman, "Why Rejection Hurts: A Common Neural Alarm System for Physical and Social Pain," *Trends in Cognitive Science* (2004), doi:10.1016/j.tics.2004.05.010.
12. N. I. Eisenberger et al., "Does Rejection Hurt? An fMRI Study of Social Exclusion," *Science* 302 (2003): 290–92.
13. C. N. DeWall et al., "Tylenol Reduces Social Pain: Behavioral and Neural Evidence," *Psychological Science* 21 (2010): 931–37.
14. For more on these ideas, see N. Eisenberger and G. Kohlrieser, "Lead with Your Heart, Not Just Your Head," *Harvard Business Review*, November 16, 2012, http://blogs.hbr.org/2012/11/are-you-getting-personal-as-a/.
15. For a review, see T. Blass, *The Man Who Shocked the World: The Life and Legacy of Stanley Milgram* (New York: Basic Books, 2009). See also T. Blass, "Understanding the Behavior in the Milgram Obedience Experiment: The Role of Personality, Situations, and Their Interactions," *Journal of Personality and Social Psychology* 60 (1991): 398–413.
16. For example, midbrain regions such as periaqueductal gray, which "control fast reflexive behaviors (e.g., fight, flight, or freeze) as well as

fear-induced analgesia." D. Mobbs et al., "When Fear is Near: Threat Imminence Elicits Prefrontal-Periaqueductal Gray Shifts in Humans," *Science* 317 (2007): 1079–83.

17. Within the psychological community, there has been a recent debate about the strength of the connection between moral judgments and physical cleanliness. I present examples of current work in this area and invite interested readers to review the overviews. H. A. Chapman and A. K. Anderson, "Things Rank and Gross in Nature: A Review and Synthesis of Moral Disgust," *Psychological Bulletin* 139 (2013): 300-27. See also B. D. Earp, J. A. C. Everett, E. N. Madva, and J. K. Hamlin, "Out, Damned Spot: Can the 'Macbeth Effect' Be Replicated?" *Basic and Applied Social Psychology* (in press).
18. For a review, see S. Lee and N. Schwartz, "Wiping the Slate Clean: Psychological Consequences of Physical Cleansing," *Current Directions in Psychological Science* 20 (2011): 307–11.
19. D. Cohen and A. Leung, "The Hard Embodiment of Culture," *European Journal of Social Psychology* 39 (2009): 1278–89.
20. A. J. Xu, R. Zwick, and N. Schwarz, "Washing Away Your (Good or Bad) Luck: Physical Cleansing Affects Risk-Taking Behavior," *Journal of Experimental Psychology: General* (2011), doi:10.1037/a0023997.
21. Though, for a great exception to this statement, readers might be interested in J. J. Ratey and E. Hagerman, *Spark: The Revolutionary New Science of Exercise and the Brain* (New York: Little, Brown and Company, 2008).

Chapter 9

1. See C. W. Cotman et al., "Exercise Builds Brain Health: Key Roles of Growth Factor Cascades and Inflammation," *Trends in Cognitive Science* (2007), doi:10.1016/j.tins.2007.06.011.
2. H. van Praag et al., "Running Increases Cell Proliferation and Neurogenesis in the Adult Mouse Dentate Gyrus," *Nature Neuroscience* 2 (1999): 266–70.
3. L. Chaddock et al., "A Neuroimaging Investigation of the Association between Aerobic Fitness, Hippocampal Volume and Memory Performance in Preadolescent Children," *Brain Research* 1358 (2010):

172–83. For related work, see L. B. Raine et al., "The Influence of Childhood Aerobic Fitness on Learning and Memory" (2013), *PloS ONE* 8: e72666, doi:10.1371/journal.pone.0072666.

4. C. H. Hillman, M. B. Pontifex, L. B. Raine, D. M. Castelli, E. E. Hall, and A. F. Kramer, "The Effect of Acute Treadmill Walking on Cognitive Control and Academic Achievement in Preadolescent Children," *Neuroscience* 159 (2009): 1044–54.
5. F. W. Booth et al., "Exercise and Gene Expression: Physiological Regulation of the Human Genome through Physical Activity," *Journal of Physiology* 543 (2002): 399–411.
6. For a review of working memory and stress effects, see S. L. Beilock, *Choke: What the Secrets of the Brain Reveal about Getting It Right When You Have To* (New York: Free Press, 2010).
7. B. A. Sibley and S. L. Beilock, "Exercise and Working Memory: An Individual Differences Investigation," *Journal of Sport and Exercise Psychology* 29 (2007): 783–91.
8. For a review of complex working memory span tasks, see A. R. A. Conway, M. J. Kane, M. F. Bunting, D. Z. Hambrick, O. Wilhelm, and R. W. Engle, "Working Memory Span Tasks: A Methodological Review and User's Guide," *Psychonomic Bulletin & Review* 12 (2005): 769–86. The problems provided are intended only to be illustrative of the types of problems one might see.
9. A. D. Brown and J. R. Curhan, "The Polarizing Effect of Arousal on Negotiation," *Psychological Science* (2013), doi:10.1177/0956797613480796. Some of the studies in this paper were not conducted with the exercise group on a treadmill but simply walking continuously, indoors or out, at a brisk pace.
10. For a nice example of Jamieson's work, see J. P. Jamieson, M. K. Nock, and W. B. Mendes, "Changing the Conceptualization of Stress in Social Anxiety Disorder: Affective and Physiological Consequences," *Clinical Psychological Science* (2013), doi:10.1177/2167702613482119.
11. As quoted in Matt Richtel, "Work Up a Sweat, and Bargain Better," *New York Times*, November 9, 2013.
12. M. Aberg et al., "Cardiovascular Fitness is Associated with Cognition in Young Adulthood," *Proceedings of the National Academy of Sciences, USA* (2009), doi:10.1073 pnas.0905307106.

13. S. J. Colcombe et al., "Cardiovascular Fitness, Cortical Plasticity, and Aging," *Proceedings of the National Academy of Sciences, USA* 101 (2004): 3316–21. This study was with physically fit older adults.
14. See D. M. Blanchette et al., "Aerobic Exercise and Cognitive Creativity: Immediate and Residual Effects," *Creativity Research Journal* 17 (2005): 257–64.
15. For a discussion of exercise and cognitive flexibility, see Y. Netz et al., "The Effect of a Single Aerobic Training Session on Cognitive Flexibility in Late Middle-Aged Adults," *International Journal of Sports Medicine* 28 (2006): 82–87.
16. See M. Bhalla and D. R. Proffitt, "Visual-Motor Recalibration in Geographical Slant Perception," *Journal of Experimental Psychology: Human Perception and Performance* 25 (1999): 1076–96; J. K. Witt et al., "The Long Road of Pain: Chronic Pain Increases Perceived Distance," *Experimental Brain Research* 192 (2008): 145–48; M. Sugovic and J. K. Witt, "An Older View on Distance Perception: Older Adults Perceive Walkable Extents as Farther," *Experimental Brain Research* 226 (2013): 383–91.
17. LaLanne's achievements are recounted in R. Goldstein, "Jack LaLanne, Founder of Modern Fitness Movement, Dies at 96," *New York Times*, January 23, 2011.
18. Details of Kotelko's life are described in Olga Kotelko, super great-grandmother with muscles of iron, Torino 2013: World Master's Games, December 28, 2012, http://www.torino2013wmg.org/news/olga-kotelko-una-super-bisnonna-dai-muscoli-d%E2%80%99acciaio?lang=en; Bruce Grierson, "The Incredible Flying Nonagenarian," *New York Times*, November 25, 2010.
19. S. Colombe and S. F. Kramer, "Fitness Effects on the Cognitive Functioning of Older Adults: A Meta-Analytic Study," *Psychological Science* 14 (2003): 125–30.
20. See C. W. Cotman and N. C. Berchtold, "Exercise: A Behavioral Intervention to Enhance Brain Health and Plasticity," *Trends in Neuroscience* 25 (2002): 295–301; C. H. Hillman, K. I. Erickson, and A. F. Kramer, "Be Smart, Exercise Your Heart: Exercise Effects on Brain and Cognition," *Nature Reviews Neuroscience* 9 (2008): 58–65.
21. K. I. Erickson et al., "The Brain-Derived Neurotrophic Factor Val66Met Polymorphism Moderates an Effect of Physical Activity," *Psychological Science* (2013), doi:10.1177/0956797613480367.

22. See A. D. Nation et al., "Stress, Exercise, and Alzheimer's Disease: A Neurovascular Pathway," *Medical Hypotheses* 76 (2011): 847–54.
23. J. C. Smith et al., "Semantic Memory Functional MRI and Cognitive Function after Exercise Intervention in Mild Cognitive Impairment," *Journal of Alzheimer's Disease* (2013), doi:10.3233/JAD-130467.
24. See Alzheimer's disease in-depth report (n.d.), *New York Times*, http://www.nytimes.com/health/guides/disease/alzheimers-disease/print.html; N. Scarmeas and Y. Stern, "Cognitive Reserve: Implications for Diagnosis and Prevention of Alzheimer's Disease," *Current Neurology and Neuroscience Reports* 4 (2004): 374–80.
25. T. Huang et al., "The Effects of Physical Activity and Exercise on Brain-Derived Neurotrophic Factor in Healthy Humans: A Review," *Scandinavian Journal of Medicine and Science in Sports* (2013), doi:10.1111/sms.12069.
26. See K. I. Erickson et al., "Exercise Training Increases Size of Hippocampus and Improves Memory," *Proceedings of the National Academy of Sciences* (2010), www.pnas.org/cgi/doi/10.1073/pnas.1015950108.
27. N. A. Mischel et al., "Physical (In)activity-Dependent Structural Plasticity in Bulbospinal Catecholaminergic Neurons of Rat Rostral Ventrolateral Medulla," *The Journal of Comparative Neurology* 522 (2014): 499–513. See also http://well.blogs.nytimes.com/2014/01/22/how-inactivity-changes-the-brain/?_php=true&_type=blogs&_php=true&_type=blogs&_r=1.
28. Pam Belluck, "Footprints to Cognitive Decline and Alzheimer's Are Seen in Gait," *New York Times*, July 17, 2012, http://www.nytimes.com/2012/07/17/health/research/signs-of-cognitive-decline-and-alzheimers-are-seen-in-gait.html?_r=0.
29. See the Global Health Care Declaration at http://exerciseismedicine.org/documents/EIMhealthcaredeclaration.pdf.

Chapter 10

1. M. A. Killingsworth and D. T. Gilbert, "A Wandering Mind Is an Unhappy Mind," *Science* (2010), www.sciencemag.org/cgi/content/full/330/6006/932/DC1.
2. J. A. Brewer et al., "Meditation Experience Is Associated with Differences in Default Mode Network Activity and Connectivity,"

Proceedings of the National Academy of Sciences, USA 108 (2011), 20254–59. This citation also encompasses the instructions below and the description of the mindfulness study that follows. See also B. K. Holzel et al., "How Does Mindfulness Meditation Work? Proposing Mechanisms of Action from Conceptual and Neural Perspectives," *Perspectives on Psychological Science* (2011), doi:10.1177/1745691611419671.

3. Of course, there is still a lot of work to be done to fully understand the efficacy of the variety of meditation practices that people employ and how such practices alter brain functioning. For a review of recent work, see *Social, Cognitive, and Affective Neuroscience* special issue on Mindfulness Neuroscience 8 (2013).
4. B. Draganski, "Changes in Grey Matter Induced by Training: Newly Honed Juggling Skills Show Up as a Transient Feature on a Brain-Imaging Scan," *Nature* 427 (2004): 311–12.
5. K. A. MacLean et al., "Intensive Meditation Training Improves Perceptual Discrimination and Sustained Attention," *Psychological Science* (2010), doi:10.1177/0956797610371339.
6. T. L. Jacobs et al., "Intensive Meditation Training, Immune Cell Telomerase Activity, and Psychological Mediators," *Psychoneuroendocrinology* (2010), doi:10.1016/j.psyneuen.2010.09.010. For related findings, see M. A. Rosenkranz et al., "A Comparison of Mindfulness-Based Stress Reduction and an Active Control in Modulation of Neurogenic Inflammation," *Brain, Behavior, and Immunity* 27 (2012): 174–84.
7. Although note that there was a relationship between the amount of time spent in daily meditation and vigilance at follow-up. The more people meditated each day, the better their performance on the vigilance tasks.
8. National Transportation Safety Board: Operational Factors/Human Performance, Group Chairman's Factual Report, December 4, 2009, http://dms.ntsb.gov/pubdms/search/document.cfm?docID=322735&docketID=48456&mkey=74940.
9. Y. Tang and M. Posner, "Attention Training and Attention State Training," *Trends in Cognitive Science* (2009), doi:10.1016/j.tics.2009.01.009.
10. Y. Tang et al., "Short-Term Meditation Induces White Matter Changes in the Anterior Cingulate," *Proceedings of the National*

Academy of Sciences, USA (2010), www.pnas.org/cgi/doi/10.1073/pnas.1011043107.

11. Y. Tang et al., "Brief Meditation Training Induces Smoking Reduction," *Proceedings of the National Academy of Sciences, USA* 11 (2013): 13971–75.
12. Y. Tang et al., "Neural Correlates of Establishing, Maintaining, and Switching Brain States," *Trends in Cognitive Science* 16 (2012): 330–37.
13. Specifically the medial prefrontal cortex. For more details, see Y. Tang et al., "Neural Correlates of Establishing, Maintaining, and Switching Brain States," *Trends in Cognitive Science* 16 (2012): 330–37.
14. Y. Tang et al., "Short-Term Meditation Training Improves Attention and Self-Regulation," *Proceedings of the National Academy of Sciences* 104 (2007): 17152–56.
15. For a review, see A. E. Hernandez and P. Li, "Age of Acquisition: Its Neural and Computational Mechanisms," *Psychological Bulletin* 133 (2007): 638–50.
16. S. Kempter, *How Muscles Learn: Teaching Violin with the Body in the Mind* (Van Nuys, CA: Alfred Music Publishing, 2003).
17. See http://alexandertechniqueknowledge.com/fmalexander.
18. For a review of the Alexander technique and music, see E. R. Valentine et al., "The Effect of Lessons in the Alexander Technique on Music Performance in High and Low Stress Situations," *Psychology of Music* (1995), doi:10.1177/0305735695232002. See also E. R. Valentine, "Alexander Technique," in A. Williamon (ed.), *Musical Excellence: Strategies and Techniques to Enhance Performance* (Oxford: Oxford University Press, 2004).
19. See http://babettemarkus.com/index.php?option=com_content&view=article&id=19&Itemid=21.
20. This introduction to the Alexander technique is taken from http://www.alexandertechnique.com/fma.htm.
21. P. Little et al., "Randomized Controlled Trial of Alexander Technique Lessons, Exercise, and Massage (ATEAM) for Chronic and Recurrent Back Pain," *British Medical Journal* (2008), doi:10.1136/bmj.a884.
22. W. M. McDonald, I. H. Richard, and M. R. DeLong, "Prevalence, Etiology, and Treatment of Depression in Parkinson's Disease," *Biological Psychiatry* 54 (2003): 363–75.
23. C. Stallibrass et al., "Randomized Controlled Trial of the Alexander

Technique for Idiopathic Parkinson's Disease," *Clinical Rehabilitation* 16 (2002): 695–708.

24. More research is needed in order to substantiate and expand the benefits of the Alexander technique to body and mind. Several neuroscientists are currently involved in projects to further understand the Alexander technique. See Henry Fagg, "The Alexander Technique and Neuroscience: Three Areas of Interest," *Statnews* 7 (2012), January 7.
25. See M. Forstmann et al., "'The Mind Is Willing, but the Flesh Is Weak': The Effects of Mind-Body Dualism on Health Behavior," *Psychological Science* 23 (2012): 1239–45.

Chapter 11

1. For a review of how stress stemming from one situation can spill over and affect performance in another, see C. Liston, B. S. McEwen, and B. J. Casey, "Psychosocial Stress Reversibly Disrupts Prefrontal Processing and Attentional Control," *Proceedings of the National Academy of Sciences, USA* 106 (2009), 912–17.
2. For a review of the incubation effect, a temporary shift away from an unsolved problem that leads to new solutions or insights, see U. N. Sio and T. C. Ormerod, "Does Incubation Enhance Problem Solving? A Meta-Analytic Review," *Psychological Bulletin* 135 (2009): 94–120.
3. M. Karlsson and L. Frank, "Awake Replay of Remote Experiences in the Hippocampus," *Nature Neuroscience* (2009), doi:10.1038/nn.2344.
4. For more on the power of nature, see R. Louv, *Last Child in the Woods. Saving Our Children from Nature Deficit Disorder* (Chapel Hill, NC: Algonquin Books, 2008).
5. F. E. Kuo and W. C. Sullivan, "Aggression and Violence in the Inner City: Effects of Environment Via Mental Fatigue," *Environment and Behavior* 33 (2001): 543–71.
6. C. M. Tennessen and B. Cimprich, "Views to Nature: Effects on Attention," *Journal of Environmental Psychology* 15 (1995): 77–85.
7. See T. Klingberg, *The Overflowing Brain* (Oxford: Oxford University Press, 2009).

8. As an example, see A. F. Taylor et al., "Coping with ADHD: The Surprising Connection to Green Play Settings," *Environment and Behavior* 33 (2001): 54–77.
9. W. James, *The Principles of Psychology* (Cambridge, MA: Harvard University Press, 1890).
10. P. Aspinall et al., "The Urban Brain: Analyzing Outdoor Physical Activity with Mobile EEG," *British Journal of Sports Medicine* (2013), doi:10.1136/bjsports-2012-091877.
11. S. Kaplan, "The Restorative Benefits of Nature: Toward an Integrative Framework," *Journal of Environmental Psychology* 15 (1995): 169–82. See also S. Kaplan and M. Berman, "Directed Attention as a Common Resource for Executive Functioning and Self-Regulation," *Perspectives on Psychological Science* 5 (2010): 43.
12. See M. G. Berman et al., "Interacting with Nature Improves Cognition and Affect in Depressed Individuals," *Journal of Affective Disorders* 140 (2012): 300–305. Mood improves after a nature walk but does not relate to memory gains seen after walking in nature. See also B. Cimprich and D. L. Ronis, "An Environmental Intervention to Restore Attention in Women with Newly Diagnosed Breast Cancer," *Cancer Nursing* 26 (2003): 284.
13. The statistics regarding city living are taken from F. Lederbogen et al., "City Living and Urban Upbringing Affect Neural Social Stress Processing in Humans," *Nature* (2011), doi:10.1038/nature10190. See this paper also for the original report of Meyer-Lindenberg's results.
14. K. C. Bickart et al., "Amygdala Volume and Social Network Size in Humans," *Nature Neuroscience* (2010), doi:10.1038/nn.2724.
15. S. Schnall et al., "Social Support and the Perception of Geographical Slant," *Journal of Experimental Social Psychology* (2008), doi:10.1016/j.jesp.2008.04.011.
16. B. P. Meier and M. D. Robinson, "Does 'Feeling Down' Mean Seeing Down? Depressive Symptoms and Vertical Selective Attention," *Journal of Research in Personality* 40 (2005): 451–61.
17. H. Davis et al., "fMRI BOLD Signal Changes in Elite Swimmers while Viewing Videos of Personal Failure," *Brain Imaging and Behavior* 2 (2008): 84–93.

Index

Page references in italics indicate illustrations.

About the Author

Sian Beilock, a leading expert on the brain science behind human performance, is a professor in the psychology department at the University of Chicago. She has PhDs in both kinesiology and psychology from Michigan State University and received an award for Transformative Early Career Contributions from the Association for Psychological Science in 2011.